ADVANCES IN TITRATION TECHNIQUES

Edited by **Vu Dang Hoang**

Advances in Titration Techniques
http://dx.doi.org/10.5772/66980
Edited by Vu Dang Hoang

Contributors

Alberto Rojas-Hernández, Norma Rodríguez-Laguna, Maria-Teresa Ramírez-Silva, Rodolfo Gómez-Balderas, Rosario Moya-Hernández, Mario A. Romero-Romo, Jean-Pierre Grolier, Jose Manuel del Río, Monica Corea, Agustín G. Asuero, Julia Martín, Lucía Fernández Sarria, Gabriel Delgado Martín, Tadeusz Michalowski, Anna Maria Michałowska-Kaczmarczyk, Aneta Spórna-Kucab

Published by InTech

Janeza Trdine 9, 51000 Rijeka, Croatia

Publishing Process Manager Edi Lipovic
Technical Editor SPi Global
Cover InTech Design team

First published September, 2017
Printed in Croatia
Legal deposit, Croatia: National and University Library in Zagreb

Additional hard copies can be obtained from orders@intechopen.com

Advances in Titration Techniques, Edited by Vu Dang Hoang
p. cm.
Print ISBN 978-953-51-3523-4
Online ISBN 978-953-51-3524-1

For EU product safety concerns:
IN TECH d.o.o., Prolaz Marije Krucifikse Kozulić 3, 51000 Rijeka, Croatia,
info@intechopen.com or visit our website at intechopen.com.

Meet the editor

Vu Dang Hoang completed his doctorate in pharmaceutics at the University of Strathclyde, UK, in 2005 and conducted a postdoctoral research at the Ecole Nationale d'Ingénieurs des Techniques des Industries Agricoles et Alimentaires, France, in 2006. He has been lecturing at the Department of Analytical Chemistry and Toxicology, Hanoi University of Pharmacy, Vietnam, since 2007. He became an associate professor in the field of drug quality control in 2015. His research interests include chemico-physical characterization of topical drug delivery systems and chemometrics-based methods for the analysis of drugs in pharmaceutical dosage forms and biological fluids. So far, he has authored more than 30 publications in pharmaceutical research and analytical chemistry in leading national and international peer-reviewed journals.

Contents

Preface

In chemistry, titration (a.k.a. titrimetry) is a common laboratory technique used for the determination of the unknown concentration of an analyte. Because of its versatility, the application of various forms of titration can affect nearly all aspects of society. This book is specifically aimed at broadening and deepening the theory and applications of titration. It contains six chapters being organized into three main sections: Volumetric Titration, Isothermal Titration Calorimetry, and Titrimetric Principles in Electrolytic Systems. Each chapter has been well written by internationally renowned experts in the field of chemistry, with mathematical expressions and illustrative examples selectively and logically presented. It is highly recommended for postgraduate students and scientists alike.

Vu Dang Hoang,
Department of Analytical Chemistry and Toxicology,
Hanoi University of Pharmacy,
Hanoi, Vietnam

Section 1

Volumetric Titration

Chapter 1

The Conditions Needed for a Buffer to Set the pH in a System

Norma Rodríguez-Laguna,
Alberto Rojas-Hernández, María T. Ramírez-Silva,
Rosario Moya-Hernández,
Rodolfo Gómez-Balderas and
Mario A. Romero-Romo

Additional information is available at the end of the chapter

http://dx.doi.org/10.5772/intechopen.69003

Abstract

It is a known fact that buffer systems are widely used in industry and diverse laboratories to maintain the pH of a system within desired limits, occasionally narrow. Hence, the aim of the present work is to study the *buffer capacity* and *buffer efficacy* in order to determine the useful conditions to impose the pH on a given system. This study is based on the electroneutrality and component balance equations for a mixture of protons polyreceptors. The added volume equations are established, V, for strong acids or bases, as well as the *buffer capacity* equations with dilution effect, β_{dil}, and the *buffer efficacy*, ε, considering that the analyte contains a mixture of the species of the same polyacid system or various polyacid systems. The ε index is introduced to define the performance of a buffer solution and find out for certain, whether the buffer is adequate to set the pH of a system, given the proper conditions and characteristics.

Keywords: buffer, buffer capacity, buffer efficacy, polyacid systems, electroneutrality equation.

1. Introduction

Currently, there are studies that examine the progress of an acid-base titration for one or various polydonor systems, extending sometimes this study to the theme of buffer capacity [1–16]. In the scientific literature, there are algorithms and simulators to construct acid-base titration curves, even considering a wide range of different mixtures of polydonor systems [17–20].

The buffer solutions have a certain buffering capacity that is used to maintain constant the pH of a system, having only a small uncertainty. The *buffer capacity, β*, has been defined as the quantity of strong acid or strong base (in the buffer solution) that gives rise to a change of one pH unit in 1 L of solution, as an intensive property of the system [15]. This involves using directly the concentration of either a strong base or an acid in the buffer solution, without considering the dilution effect, as King and Kester [2], Segurado [3], Urbansky and Schock [4], De Levie [8] did, among others. Urbansky and Schock also mentioned the use of concentration to simplify the maths. Nevertheless, the dilution effect on buffer capacity was first considered by Michałowski, as Asuero and Michałowski have established in a thorough and holistic review [6].

The buffer capacity considering the effect of dilution, β_{dil}, is defined as the added amount of strong base or strong acid required to change in one unit the pH of an initial V_o volume of the buffer solution formed by species of only one polydonor system [7]. By their definition, β is an intensive property by considering the concentration, while β_{dil} is an extensive property to include the amount of substance.

A buffer solution is used to impose the pH in a given system; generally speaking, the buffer is added to a working system selected to impose a given pH, thus giving rise to a mixture between both systems. It would be convenient to know the minimum concentration of the buffer components in the mixture as well as the minimum volume that must be added in order for it to fulfill its function. From the existing works in the literature, an evaluation of the buffer performance was attempted, in general a β was provided, although up to now this problem has not been dealt with quantitatively.

Figure 1 shows the scheme, in which a buffer solution is used to buffer the pH of a given system. It is observed that a mixture of both systems exists: the buffer system (BS) and the original system (OS). In general, it is necessary to define which would be the BS and the study system.

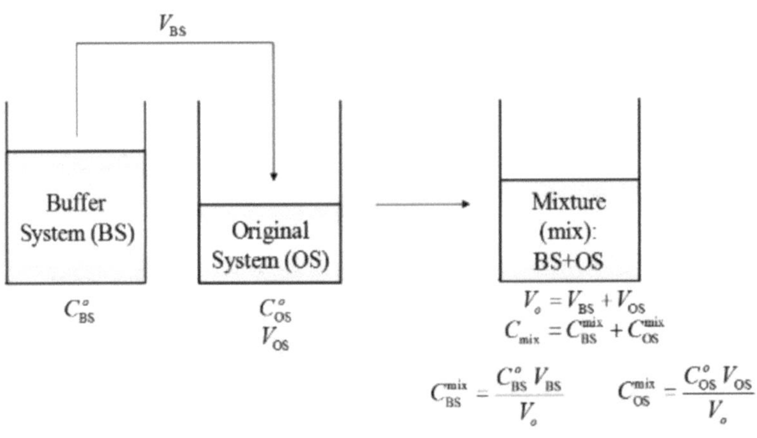

Figure 1. Schematization of the manner, in which the pH is generally set in an original system (OS) with a buffer system (BS). C^o_{BS} is the initial concentration of the buffer system before mixing, V_{BS} is the volume of BS added to OS, C^o_{OS} and V_{OS} are the initial concentration and initial volume of the original system before mixing, respectively; V_o is the mixture volume formed by the solutions with volumes V_{BS} and V_{OS}, C_{mix} is the overall concentration of the mixture, C^{mix}_{BS} is the buffer components' concentration in the mixture; and C^{mix}_{OS} is the original system's solute concentration in the mixture.

In many applications, it is convenient that the volume of the buffer solution that is added to the system of interest be $V_{BS} < V_{OS}$, in order not to alter much the composition of the study system.

The present work extends the study of β_{dil} that is contained in a mixture of polydonor systems. Moreover, a new concept is introduced, *buffer efficacy* (ε), as an index to estimate the performance of a buffer. Finally, it is shown how these indexes allow the determination of the useful conditions (minimum concentration and volume), for a buffer to enable imposing the pH in a system of interest.

2. Theoretical background

2.1. Description of the components, species, equilibria, and fractions in a mixture of polydonor systems

In order to evaluate the useful performance of a buffer to impose the pH in a given system, it is necessary to establish the expressions of β or of β_{dil} that consider mixtures of various polydonor systems.

Although Ref. [8] presented equations that describe the behavior of β for polydonor systems, the nomenclature, which has shown in Ref. [7] to study β_{dil}, is considered here to generalize its equations for the case of buffer solutions of mixtures of different polydonor systems.

A polyprotic system [6, 7] can be represented as follows:

$$H_nL^{(n-a)}/\ldots/H_jL^{(j-a)}/\ldots\ L^{a-}/H^+, \quad \text{where } j \in \{0, 1, \ldots, a, \ldots, n\} \tag{1}$$

$H_nL^{(n-a)}$ is the polyprotic acid (weak acid in general), L^{a-} is the base of the system, and the neutral species is H_aL; H^+ is the exchanged particle in the reaction, n is the number of protons of the polyprotic acid, $a-$ is the charge of the base (expressed in elementary charge units).

The species that go from $H_{(n-1)}L^{(n-a)}$ up to $HL^{(a-1)}$ are the system's formal ampholytes.

The global formation equilibria of the species of a polydonor system are represented according to Eq. (2).

$$L^{a-} + jH^+ \rightleftarrows H_jL^{(j-a)} \quad \text{with} \quad \beta_j = \frac{[H_jL^{(j-a)}]}{[L^{a-}][H^+]^j}. \qquad \text{where } j \in \{0, 1, \ldots, a, \ldots, n\} \tag{2}$$

By definition $\beta_o = 1$.

When there is a mixture of c polydonor systems in aqueous solution with $(c + 1)$ components, a general representation of the set of polyprotic systems is given as:

$$H_{n_k}\,(Lk)^{(n_k-a_k)}/\ldots/H_{j_k}\,(Lk)^{(j_k-a_k)}/\ldots/(Lk)^{a_k-}\,/H^+ \tag{3}$$

where $k \in \{1, 2, \ldots, c\}$, $j_k \in \{0, 1, \ldots, a_k, \ldots, n_k\}$.

$H_{n_k}(Lk)^{(n_k-a_k)}$ is the polyprotic acid of the kth polydonor system, $(Lk)^{a_k-}$ is its polybase, n_k is the number of protons of the kth polyprotic acid, and a_k is the charge of the kth polybase. The species that go from $H_{(n_k-1)}(Lk)^{(n_k-a_k-1)}$ up to $H(Lk)^{(1-a_k)}$ are the system's ampholytes.

A representation of the kth polydonor system's global formation equilibria in a mixture is given as:

$$(Lk)^{a_k-} + j_k H^+ \rightleftarrows H_{j_k}(Lk)^{(j_k-a_k)} \quad \text{with} \quad \beta_{j_k} = \frac{[H_{j_k}(Lk)^{(j_k-a_k)}]}{[(Lk)^{a_k-}][H^+]^{j_k}} \tag{4}$$

where $k \in \{1, 2, \ldots, c\}$, $j_k \in \{0, 1,.., a_k, \ldots, n_k\}$.

Also, in this case $\beta_{0_k} \equiv 1$ needs to be considered.

It can be demonstrated that the molar fraction to describe each of the c distributions of the species of each of the polydonor systems in the mixture with respect to H^+ is given by Eq. (5):

$$f_{j_k} = \frac{[H_{j_k}(Lk)^{(j_k-a_k)}]}{[Lk]_T} = \frac{\beta_{j_k}[H^+]^{j_k}}{\sum_{j_k=0}^{n_k} \beta_{j_k}[H^+]^{j_k}} \tag{5}$$

where $k \in \{1, 2, \ldots, c\}$, $j_k \in \{0, 1,.., a_k, \ldots, n_k\}$.

where $[Lk]_T$ is the total concentration of the kth component in the mixture. As can be observed, the molar fractions only depend on pH and on the equilibrium constants β_{j_k}.

2.2. Description of the mixture to be titrated

There are N solutions, each containing one $H_{j_k}(Lk)^{(j_k-a_k)}$ species in a C_{oj_k} molar concentration, assuming a volume V_{oj_k} is taken from each solution to form only one mixture with an overall volume V_o, then: $N = \sum_{k=1}^{c}\left\{\sum_{j_k=0}^{n_k}(1)\right\} = \sum_{k=1}^{c}\{n_k+1\}$ and $V_o = \sum_{k=1}^{c}\left\{\sum_{j_k=0}^{n_k} V_{oj_k}\right\}$. This mixture is titrated with a strong MOH base at C_b concentration or with a strong MX acid at C_a concentration, measuring the pH.

Each species has associated countercations or counteranions ($M^{z_{j_k}+}$ or $Z^{z_{j_k}-}$) depending on whether (j_k-a_k) they are negative or positive, which lack the acid-base properties.

2.3. Expressions for the titration plots of polydonor systems mixtures

Although Asuero and Michałowski [6] and De Levie [9] have presented some mathematical representations of added volume as a function of pH for these systems, we have preferred to follow the same procedure and notation used to deduce the added volume equations, proposed by Rojas-Hernández et al. [7]. Then, for the case of the mixtures of species of various polydonor systems, the following expressions can be deduced:

The added volume expression for a strong base, V_b, is

$$V_b = \frac{\sum_{k=1}^{c}\left\{\sum_{j_k=0}^{n_k}\{(j_k - a_k)(V_{oj_k}C_{oj_k})\} - \left[\sum_{j_k=0}^{n_k}(V_{oj_k}C_{oj_k})\right]\left[\sum_{j_k=0}^{n_k}\{(j_k - a_k)f_{j_k}\}\right]\right\} - V_o([\mathrm{H}^+] - [\mathrm{OH}^-])}{C_b + [\mathrm{H}^+] - [\mathrm{OH}^-]} \tag{6}$$

If one considers now that a strong acid is added to the mixture, then the V_a expression becomes:

$$V_a = \frac{-\sum_{k=1}^{c}\left\{\sum_{j_k=0}^{n_k}\{(j_k - a_k)(V_{oj_k}C_{oj_k})\} + \left[\sum_{j_k=0}^{n_k}(V_{oj_k}C_{oj_k})\right]\left[\sum_{j_k=0}^{n_k}\{(j_k - a_k)f_{j_k}\}\right]\right\} + V_o([\mathrm{H}^+] - [\mathrm{OH}^-])}{C_a - [\mathrm{H}^+] + [\mathrm{OH}^-]} \tag{7}$$

Eqs. (6) and (7), $[\mathrm{OH}^-] = \frac{K_w}{[\mathrm{H}^+]}$ agree with the water self-protolysis equilibrium.

As can be observed, the added volume equations obtained for a strong base or a strong acid bear the same mathematical form. It is relevant to note which comes from the component's balance of each polyprotic system must be independently added, thus giving rise to the double summations appearing in Eqs. (6) and (7).

When $c = 1$ hence giving $k = 1$, the equations are the same as those shown in reference [7], to determine the volume that is added to a strong base and a strong acid (V_b and V_a, respectively) in a system formed by species of the same polydonor system.

Eqs. (6) and (7) are exact analytic solutions to obtain titration plots pH = f(V) (estimating the volume from the pH values). These equations also allow obtaining exact equations of dpH/dV, hence the expressions for β_{dil} will be shown in the next section. β_{dil} is the first index used to explore quantitatively the application conditions for a buffer.

2.4. General expressions of dpH/dV_b and $-d$pH/dV_a

Eqs. (6) and (7) are functions of the pH, thus it becomes possible to obtain analytic expressions for their first derivatives (dV/dpH). With the reciprocals of the first derivatives, exact algebraic expressions of the first derivative of the titration plot are obtained. This is to say dpH/dV_b and $-d$pH/dV_a, which are used to detect the volumes at the titration points when the reactions are quantitative.

Extending the expressions for dpH/dV_b and $-d$pH/dV_a considering a mixture of the species of various polydonor systems, the expressions obtained are as follows:

$$\frac{d\mathrm{pH}}{dV_b} = \frac{C_b + 10^{-\mathrm{pH}} - 10^{\mathrm{pH}-pK_w}}{-2.303\sum_{k=1}^{c}\left(\left[\sum_{j_k=0}^{n_k}(V_{oj_k}C_{oj_k})\right]\left[\sum_{j_k=0}^{n_k}\left\{j_k f_{j_k}\sum_{i_k=0}^{n_k}\left[(i_k - j_k)f_{i_k}\right]\right\}\right]\right) + 2.303(V_o+V_b)[10^{-\mathrm{pH}}+10^{\mathrm{pH}-pK_w}]} \tag{8}$$

$$-\frac{dpH}{dV_a}=\frac{C_a-10^{-pH}+10^{pH-pK_w}}{\begin{array}{l}-2.303\sum_{k=1}^{c}\left(\left[\sum_{j_k=0}^{n_k}(V_{oj_k}C_{oj_k})\right]\left[\sum_{j_k=0}^{n_k}\left\{j_kf_{j_k}\sum_{i_k=0}^{n_k}\left[(i_k-j_k)f_{i_k}\right]\right\}\right]\right)\\+2.303(V_o+V_b)[10^{-pH}+10^{pH-pK_w}]\end{array}} \tag{9}$$

where $k\in\{1, 2, \ldots, c\}$, $i_k\in\{0, 1, \ldots, n_k\}$ and $j_k\in\{0, 1, \ldots, a_k, \ldots, n_k\}$.

Equally, if $c=1$ and $k=1$, Eqs. (8) and (9) are the same as those shown in Ref. [7] to determine dpH/dV_b and $-dpH/dV_a$ for a mixture of the species of only one polydonor system.

2.5. General expressions of β_{dil}

In order to determine the buffer capacity considering the dilution, β_{dil}, the derivative is applied to the quantity of strong base or strong acid added as follows:

$$\beta_{dil_b}=\frac{dV_bC_b}{dpH}=C_b\frac{dV_b}{dpH}\quad \text{or}\quad \beta_{dil_a}=-\frac{dV_aC_a}{dpH}=-C_a\frac{dV_a}{dpH} \tag{10}$$

where β_{dil_b} and β_{dil_a} are units of quantity of substance. The analytic mathematical expressions are shown in Eqs. (11) and (12).

$$\beta_{dil_b}=\frac{dV_bC_b}{dpH}=\frac{2.303C_b\left\{\begin{array}{l}-\sum_{k=1}^{c}\left(\left[\sum_{j_k=0}^{n_k}(V_{oj_k}C_{oj_k})\right]\left[\sum_{j_k=0}^{n_k}\left\{j_kf_{j_k}\sum_{i_k=0}^{n_k}[(i_k-j_k)f_{i_k}]\right\}\right]\right)\\+(V_o+V_b)[10^{-pH}+10^{pH-pK_w}]\end{array}\right\}}{C_b+10^{-pH}-10^{pH-pK_w}} \tag{11}$$

$$\beta_{dil_a}=-\frac{dV_aC_a}{dpH}=\frac{2.303C_a\left\{\begin{array}{l}-\sum_{k=1}^{c}\left(\left[\sum_{j_k=0}^{n_k}(V_{oj_k}C_{oj_k})\right]\left[\sum_{j_k=0}^{n_k}\left\{j_kf_{j_k}\sum_{i_k=0}^{n_k}[(i_k-j_k)f_{i_k}]\right\}\right]\right)\\+(V_o+V_b)[10^{-pH}+10^{pH-pK_w}]\end{array}\right\}}{C_a-10^{-pH}+10^{pH-pK_w}} \tag{12}$$

Furthermore, it is worth noting that the different plots presented along this work were constructed from spreadsheets done through Excel 2007 in Microsoft Office using the equations heretofore presented.

3. Some case studies

3.1. Application of expressions for the titration curves fitting experimental data of the Britton-Robinson buffer

Figure 2 shows the $pH=f(V_b)$ and curve of $dpH/dV_b=f(pH)$ retaking experimental data from *Lange's Handbook of Chemistry* [21] (markers • and ▲, respectively). The experimental curve

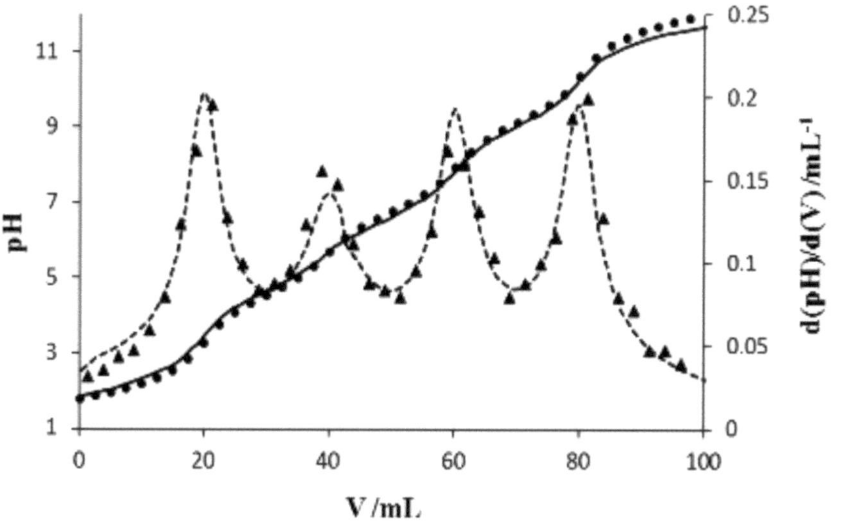

Figure 2. Titration of 100 mL of equimolar solution of Britton-Robinson 0.04 M ($[AcO'] = [PO_4'] = [BO_3'] = C_{SB} = 0.04$ M) with NaOH 0.2M.—represents the calculated curve of pH = f(V_b), • denotes the experimental curve of pH = f(V_b), ----- denotes the calculated curve of dpH/dV_b = f(V_b), and ▲ denotes the experimental curve of $\Delta pH/\Delta Vb$. The pK_a values are as follows: $pK_a = 4.66$ for acetic acid [23]; $pK_{a1} = 2.1$ [24], $pK_{a2} = 6.75$ [25], and $pK_{a3} = 11.71$ [26] for phosphoric acid; and $pK_a = 9.15$ for boric acid [27]. $pK_w = 13.73$ [28, 29].

dpH/dV_b was calculated as the finite differences quotient of the pH values and the volumes measured during the titration ($\Delta pH/\Delta V_b$) using the average volumes for each interval. Also shown are the pH = f(V_b) and dpH/dV_b = f(pH) curves obtained using Eqs. (6) and (7) (solid and segmented lines, respectively) [22].

3.2. Effect of the quantity of a buffer solution on β_{dil}

Intuitively, it is known that the performance of a buffer solution is better whenever a larger volume is taken to set the pH. The β shown in the scientific literature [1–14, 20] does not consider this feature, for which it is necessary to have an index that evaluates the effect of the size of the buffer solution to impose the pH. For that purpose, the definition of β_{dil} is used to include in its mathematic expression, the term V_o, as observed in Eqs. (11) and (12). Subsequently, it is shown how β_{dil} takes this effect into account.

Figure 3 shows a series of β_{dil} = f(pH) plots for a 1000 and 10 mL buffer solutions containing the species H_3PO_4 and $H_2PO_4^-$ at different concentrations of the PO_4' ($PO_4' = H_3PO_4/H_2PO_4^-/HPO_4^{2-}/PO_4^{3-}/H^+$) system. In this case, these concentrations can be represented as $C^o_{BS} = C_{mix} = C^{mix}_{BS}$ because the buffer system is a mixture, in agreement with **Figure 1**. It is necessary to underline that the axis β_{dil} is log, just as Urbansky and Schok [4] do, in order to compare β_{dil} within an ample PO_4' concentration range.

Figure 3a represents a larger buffer system than that represented in **Figure 3b**, because the initial volumes were 1000 and 10 mL, respectively. As can be observed in **Figure 3**, β_{dil} increases with

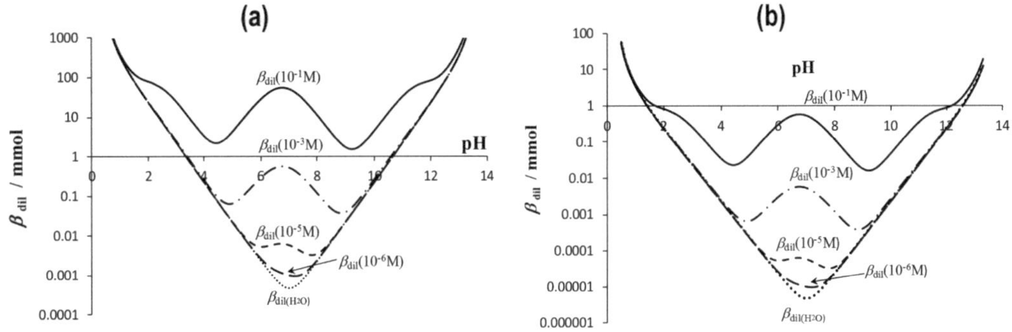

Figure 3. Calculated plots of $\beta_{dil} = f(\mathrm{pH})$ of systems initially containing the H_3PO_4 and $H_2PO_4^-$ at PO_4' species at overall concentrations of 10^{-1}, 10^{-3}, 10^{-5} and 10^{-6} M. The broken line represents the plot of $\beta_{dil(H2O)} = f(\mathrm{pH})$ for water and its basic and acid particles. $C_b = C_a = 0.5$ M. $pK_{a1} = 2.1$ [24], $pK_{a2} = 6.75$ [25], and $pK_{a3} = 11.71$ [26]. (a) $V_o = 1000$ mL. (b) $V_o = 10$ mL.

increasing quantity of the system (V_o). Hence, β_{dil} indicates well the expected behavior for a buffer system: the pH in a system is better imposed when the buffer amount is larger.

Figure 3 also shows the plot for water, $\beta_{dil(H2O)} = f(\mathrm{pH})$, titrated with strong base and strong acid. It sets the lower limit given by this solvent and its acid and basic particles (broken line), with respect to all aqueous solutions. Therefore, it is established that any solution, including that of the same solvent, has certain β_{dil}. It can be observed that the concentration diminution of the PO_4' system provokes that β_{dil} diminishes and the width of the pH interval also decreases where the PO_4' system contributes more to β_{dil} than the solvent.

There is one *minimum concentration of the buffer system* (C_{BSmin}), small enough, where the PO_4' system almost does not contribute to β_{dil}, ($C_{BSmin} \approx 10^{-6}$ M), so that the plot of β_{dil} of the PO_4' system can be discerned from the plot of $\beta_{dil(H2O)}$. Just as a minimum concentration is shown for the PO_4' system; whenever there is an acid-base pair (HL/L) with $pK_a \approx 7.0$, the C_{BSmin} will be the same (10^{-6} M). Although in other cases, when $pK_a < 7.0$ or $pK_a > 7.0$, it must be expected that the C_{BSmin} be larger; this is to say, for $pK_a = 5$ or $pK_a = 9$, the $C_{BSmin} \approx 10^{-5}$ M and for $pK_a = 3$ or $pK_a = 11$, the $C_{BSmin} \approx 10^{-3}$ M.

3.3. Buffer efficacy (ε) of a buffer system

The previous section showed that the β_{dil} has advantages over β in order to evaluate the buffer performance. However, the shortcomings of this situation refer to β_{dil}, which by definition is the quantity of strong base or strong acid added to change by one unit the system's pH, which is a fairly large change. Therefore, it is necessary to define a new index having a smaller change than β_{dil}.

Following the idea proposed by Christian in his textbook [30], it is possible to approximate the derivative by means of a finite difference quotient. Even when the pH change is acceptable for a buffer system, it depends on the application or on the system to be considered, a $\Delta\mathrm{pH} \le 0.1$ is sufficiently small to comply with the approximation established through Eq. (13).

$$\beta_{\text{dil}} = \frac{d(VC)}{d\text{pH}} \approx \frac{\Delta(VC)}{\Delta\text{pH}} \tag{13}$$

Then

$$\Delta(VC) \approx \beta_{\text{dil}} \Delta\text{pH} \tag{14}$$

In this work, the buffer performance will be assessed considering a $\Delta\text{pH} = 0.1$ [30]. The *buffer efficacy, ε*, is defined as the quantity of strong base or strong acid that provokes a pH change of only one-tenth in a system. The expression of ε is as follows:

$$\varepsilon \equiv \Delta(VC) \approx \beta_{\text{dil}} \Delta\text{pH} \ = 0.1\beta_{\text{dil}} \tag{15}$$

3.4. Application of ε: buffer system's concentration threshold

A buffer system is used to set the pH, therefore, it is necessary to know its useful conditions to fulfill its function. Then, for the sake of a deeper understanding it is relevant to establish first a limit to determine the moment in which the buffer system's concentration sets pH conditions over those of the water and of its acid and basic particles, just as those of the system of interest.

3.4.1. *Imposing the pH of the buffer system over the water and its acid and basic particles*

Because the system's pH needs to be imposed, the efficacy of the system's buffer, the buffer (ε_{BS}) should be larger or at least equal to the *efficacy of the buffer amplified ten times that of the water*, $\varepsilon_{10(H_2O)} = 10\varepsilon_{(H_2O)}$, not just the *buffer efficacy of the water*, $\varepsilon_{(H_2O)}$s. In the present work, a factor of 10 is considered sufficiently large to assess a buffer's performance.

Figures 4a and **b** show the curves of $\varepsilon_{(H_2O)}$ for $V_o = 1000$ mL and $V_o = 10$ mL of water, respectively, titrated with a strong acid and a strong base at 1 M concentration (marker…). Also, it is shown the plot of $\varepsilon_{10(H_2O)}$ (marker…), which will be considered as the limit where the buffer concentration is useful to set the pH. It is also observed in **Figures 4a** and **b** that both $\varepsilon_{(H_2O)}$ and $\varepsilon_{10(H_2O)}$ depend, as expected, on the quantity of the system.

Figure 4c and **d** shows, apart from the $\varepsilon_{(H_2O)}$ and $\varepsilon_{10(H_2O)}$ plots, those of ε_{BS} of the NH_4^+/NH_3 buffer solutions at different concentrations for systems with $V_{BS} = V_o = 1000$ mL and $V_{BS} = V_o = 10$ mL, respectively, as those shown in **Figure 1**. In order to establish the limit in which a buffer works to set the pH, it is necessary to compare the ε_{BS} curve with that of the $\varepsilon_{10(H_2O)}$, so that compliance with $\varepsilon_{BS} \geq \varepsilon_{10(H_2O)}$ can be verified.

In this case, the lowest NH_4^+/NH_3 buffer concentration falls to a point that almost equals the $\varepsilon_{10(H_2O)}$ plot. Therefore, this concentration is termed as *threshold buffer concentration* (TC_{BS}) with a value of $TC_{BS} \approx 10^{-4}$ M. It can also be seen from **Figures 4c** and **d** that TC_{BS} does not depend on the size of the system (V_o).

Then, whenever there is a buffer formed by an acid-base pair HL/L with $pK_a \approx 7.0$ it must be expected a greater TC_{BS}, of approximately 10^{-5} M; when $pK_a = 5$ or $pK_a = 9$, the TC_{BS} will also be approximately 10^{-4} M; and for $pK_a = 3.0$ or $pK_a = 11.0$, the $TC_{BS} \approx 10^{-2}$ M.

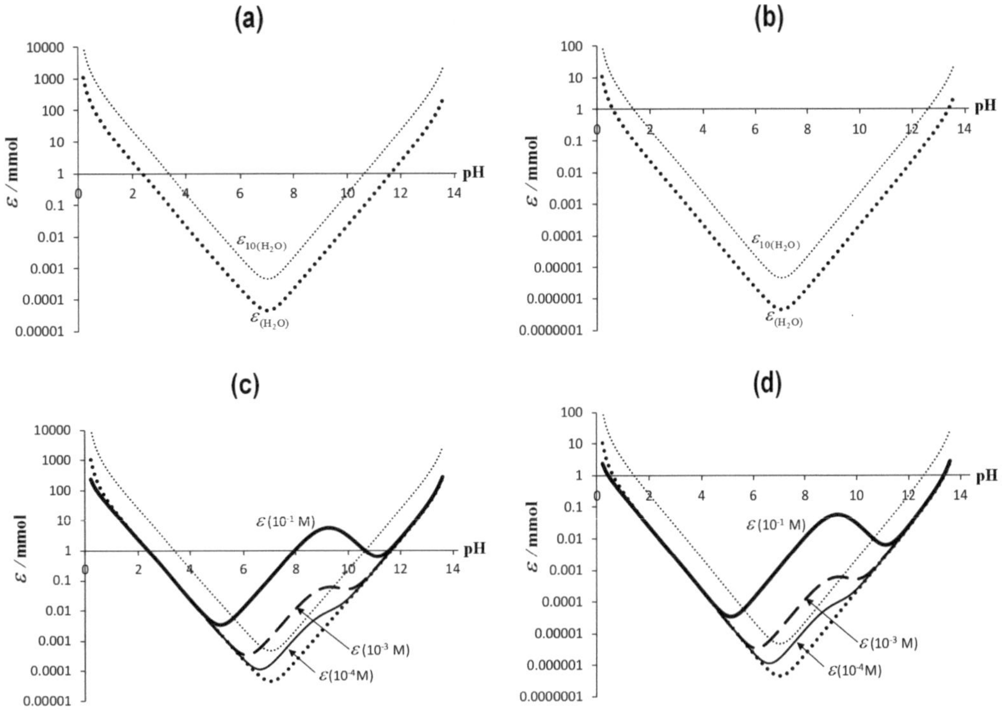

Figure 4. Calculated plots of $\varepsilon_{(H_2O)}$ and $\varepsilon_{10(H_2O)}$ for $V_o = 1000$ mL and $V_o = 10$ mL of water and its acid and basic particles (marker… and marker…, respectively). The ε_{BS} curves of NH_4^+/NH_3 buffer solutions at different concentrations, $[NH_3']_{Tot} = C^o_{BS} = C_{mix} = C^{mix}_{BS}$: -10^{-1} M, $---10^{-3}$ M, -10^{-5} M. $C_b = C_a = 1$ M. $pK_a = 9.25$ [31]. (a) and (c) 1000 mL. (b) and (d) 10 mL.

Furthermore, in **Figures 4c** and **d** it can also be noted that the useful interval for the BS to set the pH, when the $C^o_{BS} = C_{mix} = C^{mix}_{BS} = 10^{-1}$ M, is $pK_a - 1 < pH < pK_a + 1$ [32]. Whereas for $C^o_{BS} = C_{mix} = C^{mix}_{BS} = 10^{-3}$ M, the interval to set the pH is smaller because the ε_{BS} becomes closer to $\varepsilon_{10(H_2O)}$.

When the $\varepsilon_{BS} < \varepsilon_{10(H_2O)}$ strong acid or strong base must be used also at adequate concentration to set the pH. The strong base and strong acid in these extremes are used because the acid and basic particles of the solvent contribute more to ε than the buffer system components.

Finally, from the analysis of the plots in **Figure 4**, it can be determined that the pH limits of the buffer performance and the buffer system threshold concentration do not depend on the system's size.

3.4.2. *Setting the buffer's system pH over that of the original system*

It is necessary to set new limits (of pH and buffer concentration) whenever there is a mixture of the system of interest and the buffer system, because it is not sufficient to consider only the water effect.

The pH of a given system (OS) is set upon adding a buffer system (BS), hence, the buffer component concentration in the mixture (C^{mix}_{BS}) must be greater than the concentration of the

http://dx.doi.org/10.5772/intechopen.69003

solutes of the original system (C_{OS}^{mix}), because both systems have their own ε, though how large is it? (**Figure 1**).

An example of the use of ε, to evaluate setting the pH to a 9.0 value at 100 mL (V_{OS}) of an acetylacetone solution (acac') (OS), is given next, at 10^{-3} M concentration (C_{OS1}^{o}), at different NH_4^+/NH_3 buffer concentrations in the mixture (C_{BS}^{mix}).

Figure 5a shows the ε of the acetylacetone solution (OS), stated in the previous paragraph (ε_{OS}, marker ×××××), the *buffer efficacy amplified 10 times that of the original system* (ε_{10OS}, marker ----), also presenting the plot of $\varepsilon_{(H_2O)}$ (marker…).

Figure 5b shows, apart from the $\varepsilon_{(H_2O)}$ curve, those corresponding to ε_{BS} for 100 mL of the NH_4^+/NH_3 system solution at different concentrations. The ε_{BS} magnitude depends on buffer's concentration and decreases until it reaches a TC_{BS}, which is given when the ε_{BS} curve almost becomes equal to that of $\varepsilon_{10(H_2O)}$. In this case, $TC_{BS} \approx 10^{-4}$ M.

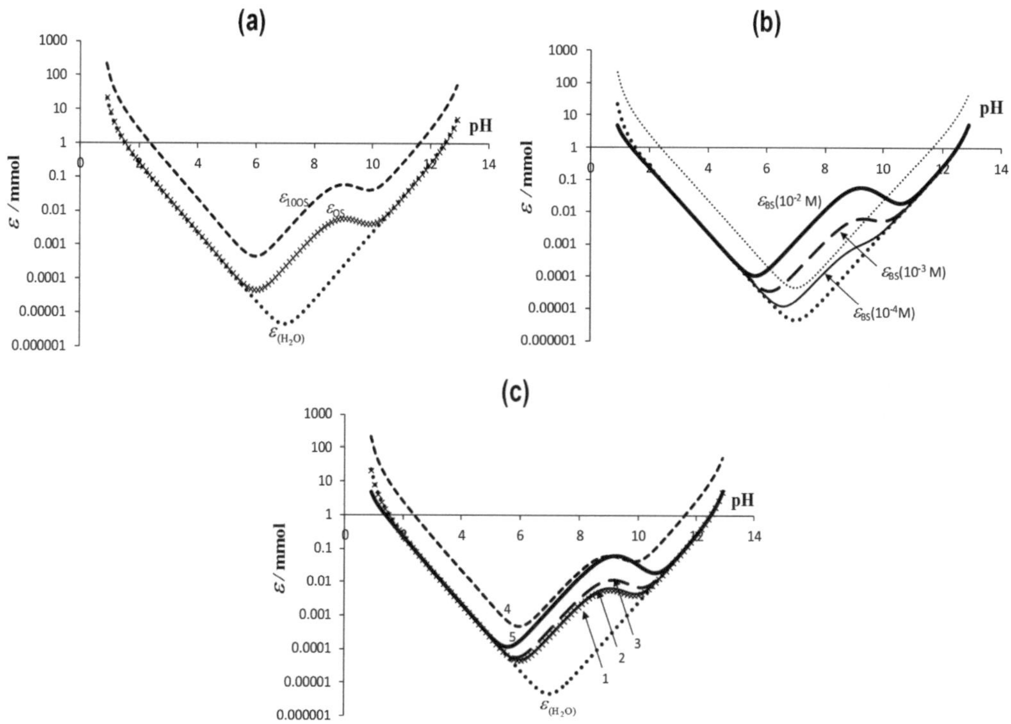

Figure 5. Curves of $\varepsilon = f(pH)$ for 100 mL solutions related to **Figure 1**. The line marked with ….in all cases represents the buffer efficacy of water, $\varepsilon_{(H_2O)}$, and its acid and basic particles. (a) Curves ε_{OS} and ε_{10OS} belong to an acac 10^{-3} M ($[acac']_{Tot} = C_{OS}^{o} = C_{mix} = C_{OS}^{mix}$) solution in the absence of the buffer system (markers ××××× and ----, respectively). (b) Curves ε_{BS} belong to the NH_4^+/NH_3 buffer at different concentrations, $[NH_3']_{Tot} = C_{BS}^{o} = C_{mix} = C_{BS}^{mix}$: —$10^{-2}$ M, – – –10^{-3} M, —10^{-4} M. (c) Curves ε_{mix}, ε_{OS}, and ε_{10OS} belong to the solutions containing acac 10^{-3} M plus NH_4^+/NH_3 10^{-2} M buffer, 10^{-3} M, and 10^{-4} M in the mixture. 1: ε_{OS}, 2:C_{OS}^{mix}=10^{-3} M +$C_{BS}^{mix} = 10^{-4}$ M, 3: $C_{OS}^{mix} = 10^{-3}$ M +$C_{BS}^{mix} = 10^{-3}$ M, 4:ε_{10OS}, and 5: C_{OS}^{mix}=10^{-3} M + $C_{BS}^{mix} = 10^{-2}$ M. $pK_a = 9.0$ for the system Hacac/acac$^-$ and $pK_a = 9.25$ for the NH_4^+/NH_3 system [31]. $C_b = C_a = 0.2$ M.

In order to impose the pH of the original system, the buffer efficacy of the mixture (ε_{mix}) should be greater or, at least equal to ε_{10OS}. Now, if the buffer is added to the original system to set the pH, the ε_{mix} has contributions of the original system and to the buffer system; thus, the buffer does not always set the pH in the system as shown in **Figure 5c**. The curve 1 that represents a ε_{OS} is practically identical to the curve 2 that corresponds to a mixture of the original system, with the buffer solution with concentration 10 times smaller than the solutes in the original system: in these cases, the pH of the system depends only on the original system, because $\varepsilon_{mix} \approx \varepsilon_{OS}$. The curve 3 shows that the buffer with 10^{-3} M concentration does already contribute to ε_{mix} apart from the original system, but has not set the pH yet because $\varepsilon_{OS} < \varepsilon_{mix} < \varepsilon_{10OS}$.

Finally, **Figure 5c** depicts the curve 4 as corresponding to ε_{10OS}, whereas the curve 5 represents a ε_{mix} corresponding to the *buffer threshold concentration for the buffer system component of the mixture*, TC_{BS}^{mix} (whenever it is required to set the pH at a value of 9.0), for which $\varepsilon_{mix} \approx \varepsilon_{BS} = \varepsilon_{10OS}$, and therefore, there is an adequate buffer performance to set the pH of the system within the $8.25 < pH < 10.0$ interval. It must be noted that if it is required to set $pH > 10.0$ values, a strong base must be used, apart from the buffer system. To the extent that $\varepsilon_{mix} >> \varepsilon_{10OS}$ the buffer performance to set the system pH becomes better. Approximately TC_{BS}^{mix} must be 10 times larger than C_{OS}^{mix}. Therefore, it is established that

$$C_{BS}^{mix} \geq TC_{BS}^{mix} \approx 10C_{OS}^{mix} \tag{16}$$

Eq. (16) becomes specific whenever the original system is set to a $pH = 9.0$ when this system bears one acid-base pair with $pK_a = 9.0$ (Hacac/acac$^-$) [31]. This example is the most difficult case because this OS acid-base pair competes almost equally with the BS acid-base pair (NH_4^+/ NH_3, $pK_a = 9.25$) to set the system's pH.

If the pK_a of some species in the original system moves away from the pH value that is desired to impose with the buffer system, the factor of 10 in Eq. (16) becomes smaller.

It must not be forgotten that the buffer system should have a conjugated acid-base pair with a pK_a value close to the pH that is desired to impose. As can be observed from **Figure 5c**, in this case it is difficult that the buffer system imposes the pH to a 9.0 value because both systems (BS and OS) have pK_a values similar to that pH value. It is worth clarifying that the NH_4^+/NH_3 buffer imposes more easily the pH to a 9.0 value to the extent that the pK_a of the acid-base pair original system drifts apart from $pH = 9.0$; of the TC_{BS}^{mix} diminishes till it reaches a limit value given by $\varepsilon_{10(H_2O)}$.

Consider now the case that a pH 9.0 shall be imposed to the 100 mL (V_{OS}) of the acetylacetone solution (acac') (OS), with a 10^{-3} M (C_{OS}^{o}) concentration, using now the Britton-Robinson [12, 21] buffer at different concentrations. For this example, the ε_{OS}, ε_{10OS}, and $\varepsilon_{(H_2O)}$ are the same as those presented in **Figure 5a**.

Figure 6a shows, apart from the curve of $\varepsilon_{(H_2O)}$, the curves of ε_{BS} for 100 mL solution of the Britton-Robinson buffer at different concentrations. In this case, the $C_{BS}^{o} = C_{mix} = C_{BS}^{mix}$ diminishes till reaching a TC_{BS}, having a value of $TC_{BS} \approx 10^{-5}$ M because at this point the ε_{BS} almost equals the $\varepsilon_{10(H_2O)}$ curve.

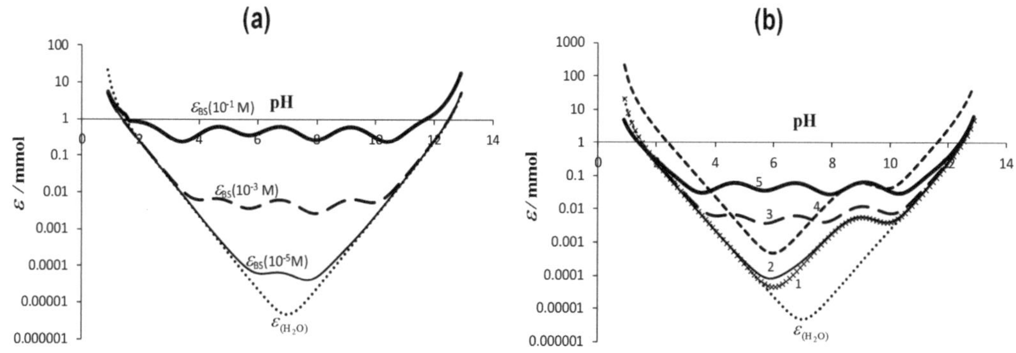

Figure 6. The curves $\varepsilon = f(\text{pH})$ are for the 100 mL of the solutions related to **Figure 1**. The broken line in all cases represents the buffer efficacy of the water, $\varepsilon_{(H_2O)}$, and its acid and basic particles. (a) Curves ε_{BS} for equimolar Britton-Robinson buffer solutions at different concentrations, $[AcO']_{Tot} = [PO_4']_{Tot} = [BO_3']_{Tot} = C^{o}_{BS} = C^{mix}_{BS}$: —$10^{-1}$ M, – – – 10^{-3} M, —10^{-5} M. (b) Curves ε_{mix}, ε_{OS}, and ε_{10OS} of solutions that contain 10^{-3} M acac plus 10^{-2} M, 10^{-3} M, and 10^{-5} M Britton-Robinson's buffer in the mixture. 1: ε_{OS}, 2: $[acac'] = C^{mix}_{OS} = 10^{-3}$ M + $C^{mix}_{BS} = 10^{-5}$ M, 3: $[acac'] = C^{mix}_{OS} = 10^{-3}$ M + $C^{mix}_{BS} = 10^{-3}$ M, 4: ε_{10OS}, and 5: $[acac'] = C^{mix}_{OS} = 10^{-3}$ M + $C^{mix}_{BS} = 10^{-2}$ M. The pK_a values used in the model are as follows: $pK_a = 9.0$ for acac′ [31]; $pK_a = 4.66$ for acetic acid [23]; $pK_{a1} = 2.1$ [24], $pK_{a2} = 6.75$ [25], and $pK_{a3} = 11.71$ [26] for phosphoric acid; and $pK_a = 9.15$ for boric acid [27]. $pK_w = 13.73$ [28, 29]. $C_b = C_a = 0.2$ M.

As stated, the ε_{mix} should be larger or at least equal to ε_{10OS}. Once again, the buffer efficacy of the mixture (ε_{mix}) is contributed from both the acac′ system and the Britton-Robinson buffer; thus the buffer does not always set the pH in the system as shown in **Figure 6b**. Curve 1 represents the ε_{OS}, that is practically equal to curve 2 corresponding to a mixture of the original system with the buffer solution with a concentration 100 times smaller than the solutes in the original system: in these cases, the system's pH depends only on the original system, because $\varepsilon_{mix} \approx \varepsilon_{OS}$. Curve 3 shows that in the $4.5 < pH < 7.4$ interval, the buffer with 10^{-3} M concentration contributes more to the ε_{mix} than the original system, consequently, the buffer has the system's capacity to set the pH in this interval but not at pH = 9.0 as required in this example. Curve 3 also has another region where both systems contribute almost the same as ε_{mix} ($7.4 < pH < 10.7$) and because $\varepsilon_{mix} < \varepsilon_{10OS}$, the buffer is not capable of fulfilling its function in this interval.

Figure 6b also shows curve 4 that corresponds to ε_{10OS} and curve 5 that represents a ε_{mix} corresponding to a *threshold concentration for the buffer system's components in the mixture*, TC^{mix}_{BS} (when the pH to be imposed is 9.0), for which $\varepsilon_{mix} \approx \varepsilon_{BS} = \varepsilon_{10OS}$. At this same concentration, the buffer system's components show a good performance to impose the pH, not only at 9.0 but within the $3.8 < pH < 10.0$ interval; which is large because it corresponds to a wide spectrum buffer. Approximately TC^{mix}_{BS} must be 10 times larger than the C^{mix}_{OS} as established by Eq. (16), because the acid-base pair of the Britton-Robinson buffer is the $H_3BO_3/H_2BO_3^-$ with a $pK_a = 9.15$ value. It must be underlined that if it is intended to impose the pH to other value ($pH < 8.1$), the TC^{mix}_{BS} is smaller.

3.5. Application of ε: threshold volume of the buffer system

The concentration of the buffer components in the mixture (C^{mix}_{BS}) must be larger than the solute concentration in the system considered (C^{mix}_{OS}) to set the pH of the system. However, it is also necessary to know the buffer minimum volume that must be added to the original system with the aim of fulfilling its function (**Figure 1**). Next, an example is given to determine this minimum volume on the acetylacetone case using Britton-Robinson buffer to impose pH 9.0 in the system.

If there are 100 mL of an acac (OS_1) solution at 10^{-3} M concentration (C^o_{OS1}), and two buffer (BS_1 y BS_2) solutions with components 10^{-1} M (C^o_{BS1}) and 1 M (C^o_{OS2}) concentration: which is the minimum volume that should be added to each buffer solution to set a pH = 9.0 in the system?

Figure 7a shows that the ε_{10OS} can be attained when 10 mL of the buffer solution with 10^{-1} M (SB_1) component concentration is added to the original system. Then, this minimum volume is the *threshold volume* (TV_{BS1}) for this specific buffer solution. If now, the buffer solution of 1 M (SB_2) component concentration is added to the system of interest, the threshold volume is different, as shown in **Figure 7b**, under these conditions the $TV_{BS2} = 1$ mL.

The TV_{BS} is related to TC^{mix}_{BS} in the mixtures, with total volumes of 110 mL for the first case and 101 mL for the second case, because in both cases TC^{mix}_{BS} are 10 times greater than the original system's solute concentrations in the mixture (C^{mix}_{OS}).

If now, a larger concentration of acac is used in the original system, for example 10^{-2} M (C^o_{OS2}), it is necessary to use the more concentrated buffer to impose the pH, for example SB_2, with 1 M (C^o_{BS2}) component concentration.

It is clear that if the original system solutes' concentrations in the mixture (C^{mix}_{OS}) grow, the TC^{mix}_{BS} also grows, consequently TV_{BS} does it too.

Observing the equations shown in **Figure 1**, it is possible to demonstrate that the threshold volume, TV_{BS}, can be determined from the initial working conditions using Eq. (15)

$$C^{mix}_{BS} \geq TC^{mix}_{BS} \approx 10C^{mix}_{OS} \tag{19}$$

which can be rewritten as

$$\frac{C^o_{BS}\,V_{BS}}{V_o} \geq \frac{C^o_{BS}\,TV_{BS}}{V_o} = 10\frac{C^o_{OS}\,V_{OS}}{V_o} \tag{17}$$

From Eq. (17), the following is obtained

$$V_{BS} \geq TV_{BS} = 10\left(\frac{C^o_{OS}V_{OS}}{C^o_{BS}}\right) \tag{18}$$

Figure 7c shows that the TV_{BS} is equal to 10 mL, even if it is a volume added to the buffer solution with 1 M component concentration, because the original system is composed of 100 mL with 10^{-2} M solute concentration (C^o_{OS2}). It can be proved that the TC_{BS} fulfills the condition of being 10 times larger than the original system's solute concentration.

http://dx.doi.org/10.5772/intechopen.69003

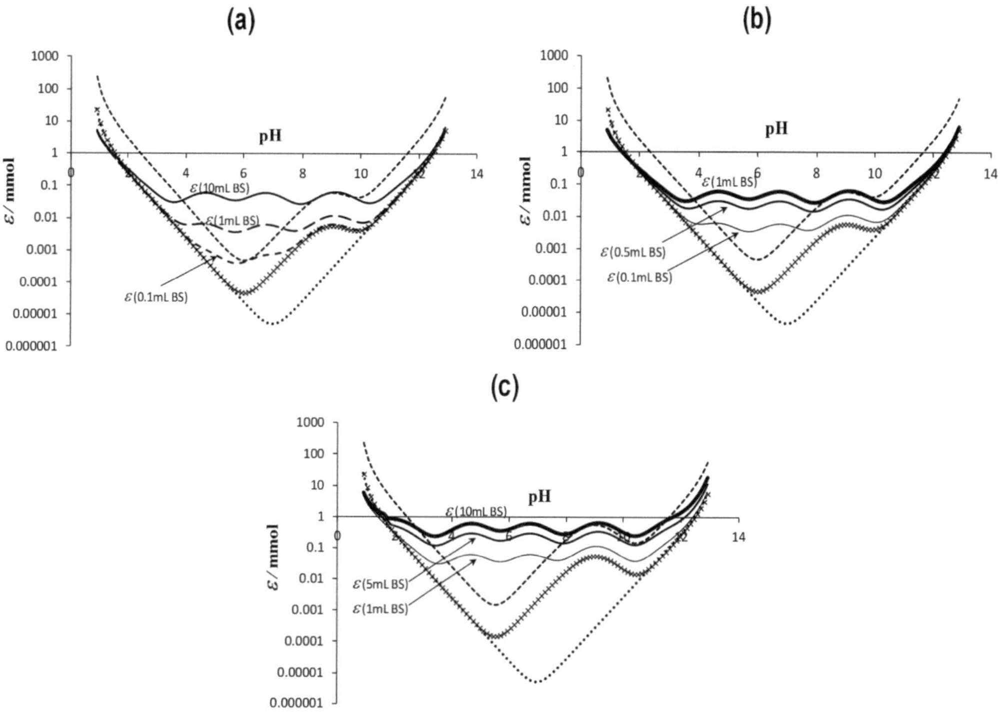

Figure 7. The curves for ε= f(pH) for 100 mL of different solutions. The broken line in all cases represents the buffer efficacy of water, $\varepsilon_{(H_2O)}$, and its acid and basic particles. The curves of ε_{OS} and ε_{100S} of acac solutions in the absence of the buffer system (markers ××××× and ---, respectively). (a) Curves ε_{OS} and ε_{100S} containing 100 mL of the acac solution with a 10^{-3} M solute concentration, curve ε_{mix} containing the same acac solution as ε_{OS} and different added volumes of the buffer solution with a concentration C_{BS} = 0.1 M. Volumes added: —10 mL, – – –1 mL and, 0.1 mL. (b) The curves ε_{OS} and ε_{100S} are the same that in (a), curves ε_{mix} containing the same acac solution and different added volumes of the buffer solution with C_{BS}= 1 M. Added volumes: —1 mL, —0.5 mL and —0.1 mL. (c) Curves ε_{OS} and ε_{100S} containing 100 mL of a acac solution with 10^{-2} M solute concentration, curves ε_{mix} containing 100 mL of the acac solution with a 10^{-2} M solute concentration and different volumes added to the buffer solution with C_{BS}= 1 M. Added volumes: —10 mL, —5 mL and, — 1 mL.

Just like in the case of the threshold concentration (TC_{BS}^{mix}) of Eq. (16), Eq. (18) also corresponds to the most difficult case. The factor of 10 can be smaller to the extent that the pK_a values of the original system move away from the pH that is to be set.

4. Conclusions

A model has been proposed to study the buffer capacity with dilution effect, β_{dil}, of mixtures of various polydonor systems in aqueous solutions. From the model, exact algebraic expressions were obtained that describe the buffer capacity with dilution.

Through the study of β_{dil} of solutions containing one or more polydonor systems with different conditions and characteristics, it is concluded that β_{dil} decreases when the total concentration of the polydonor systems (or mixtures of polydonor systems) decreases, and conversely.

When a polydonor system attains a very small concentration, this does not contributes practically to the buffer capacity, only the solvent's acid and basic particles determine this property. Then, it is stated that any solution, even the pure solvent, has a buffer capacity.

It is shown that the β_{dil} depends on the size of the system, which is information that is not considered in the β known and used in the common scientific literature.

A new index has been introduced, ε, to measure the buffer efficacy of a buffer solution such that the quantity added of strong base or strong acid causes a change of only one-tenth of a pH unit instead of one unit as β_{dil}.

From the construction of the different curves of $\varepsilon = \mathrm{f(pH)}$, it is possible to identify a buffer threshold concentration in the mixture (TC_{BS}^{mix}), which allows knowing the minimum buffer concentration to set the desired pH of the system of interest. This concentration TC_{BS}^{mix} must be, at least 10 times greater than that of the original system's solutes in the mixture ($TC_{BS}^{mix} = C_{OS}^{mix}$) in the most difficult case.

Similarly, the different curves of $\varepsilon = \mathrm{f(pH)}$ also allow determining the minimum volume (buffer threshold volume, TV_{BS}) that must be added to the system of interest to set its pH $\left(TV_{BS} = 10\left(\frac{C_{OS}^{o} V_{OS}}{C_{BS}^{o}}\right)\right)$ in the most difficult case.

Acknowledgements

NR-L wants to acknowledge DGAPA-UNAM for a postdoctoral fellowship. The authors also want to acknowledge for partial financial support from PRODEP-SEP through RedNIQAE and CONACyT through 237997 project.

Author details

Norma Rodríguez-Laguna[1,2], Alberto Rojas-Hernández[1]*, María T. Ramírez-Silva[1], Rosario Moya-Hernández[2], Rodolfo Gómez-Balderas[2] and Mario A. Romero-Romo[3]

*Address all correspondence to: suemi918@xanum.uam.mx

1 Departamento de Química, Universidad Autónoma Metropolitana-Iztapalapa, Área de Química Analítica, México, DF, Mexico

2 Departamento de Materiales, Universidad Autónoma Metropolitana-Azcapotzalco, Área de Ingeniería de Materiales, México, DF, Mexico

3 Facultad de Estudios Superiores-Cuautitlán, Universidad Nacional Autónoma de México, Estado de México, Mexico

References

[1] Van-Slyke DD. On the measurement of buffer values and on the relationship of buffer value to the dissociation constant of the buffer and the concentration and Reaction of the buffer solution. The Journal of Biological Chemistry. 1922;**52**: 525–570

[2] King DW, Kester DR. A general approach for calculating polyprotic acid speciation and buffer capacity. Journal of Chemical Education. 1990;**67**:932–933. DOI: 10.1021/ed067p932

[3] Segurado MAP. Extreme values in chemistry: Buffer capacity. The Chemical Educator. 2003;**8**:22–27

[4] Urbansky ET, Schock MR. Understanding, deriving, and computing buffer capacity. Journal of Chemical Education. 2000;**77**:1640–1644

[5] Moya-Hernández R, Rueda-Jackson JC, Ramírez MT, Vázquez GA, Havel J, Rojas-Hernández A. Statistical study of distribution diagrams for two-component systems: Relationship of means and variances of the discrete variable distributions with average ligand number and intrinsic buffer capacity. Journal of Chemical Education. 2002;**79**:389–392

[6] Asuero A, Michałowski T. Comprehensive formulation of titration curves for complex acid-base systems and its analytical implications. Critical Reviews in Analytical Chemistry. 2011;**41**:151–187

[7] Rojas-Hernández A, Rodríguez-Laguna N, Ramírez-Silva MT, Moya-Hernández R. Distribution diagrams and graphical methods to determine or to use the stoichiometric coefficients of acid-base and complexation reactions. In: Innocenti A, editor. Stoichiometry and Research - The Importance of Quantity in Biomedicine. Rijeka: InTech; 2012. pp. 287–310

[8] De Levie R. Explicit expressions of the general form of the titration curve in terms of concentration. Journal of Chemical Education. 1993; **70**:209–217

[9] De Levie R. General expressions for acid-base titrations of arbitrary mixtures. Analytical Chemistry. 1996;**68**:585–590

[10] De Levie R. Ratio diagrams: A simple graphical representation of complicated equilibria. Journal of Chemical Education. 1970;**47**:187–192

[11] Chiriac V, Balea G. Buffer index and buffer capacity for a simple buffer solution. Journal of Chemical Education. 1997;**74**:937–980

[12] Asuero AG. Buffer capacity of a polyprotic acid: first derivate of the buffer capacity and pK_a values of single and overlapping equilibria. Critical Reviews in Analytical Chemistry. 2007;**37**:269–301

[13] Janjić TJ, Pfendt LB, Pasulj MB. Two-phase buffer systems containing a sparingly soluble diprotic acid as the solid phase. Monatshefte für Chemie, Chemical Monthly. 1984;**115**: 705–711

[14] Fishtik I, Povar I. Buffer capacity in multiple chemical reaction systems involving solid phases. Canadian Journal of Chemistry. 2006;**84**:1036–1044

[15] Skoog AD, West DM, Holler JF, Crouch SR. Analytical Chemistry. Southbank: Thomson; 2004. p. 992

[16] Högfeldt E. Graphic Presentation of Equilibrium Data. Treatise on Analytical Chemistry. In: Kolthoff IM, Elving PJ, editors. Part I, Vol II, Chap.15, New York: Interscience, 1984

[17] Fleck GM. Equilibria in Solution. New York: Holt; 1966. p. 217

[18] Gutz IGR. Simulator: CurTipot. Versión 3.5.4, MS Excel. Instituto de Química de la Universidad de Sao Paulo [Internet]. 2014. Available from: http://www2.iq.usp.br/docente/gutz/Curtipot_.html [Accessed: 03 March 2017]

[19] Tarapčik P, Beinrohr E. Implementation of a Universal Algorithm for pH Calculation into Spreadsheet and its Use in Teaching in Analytical Chemistry. Slovakia: Faculty of Chemical and Food Technology STU; 2005

[20] Efstathiou CE. Acid-Base Titration Curves. Department of Chemistry, University of Athens, Greece [Internet]. 2000. Available from: http://195.134.76.37/applets/AppletTitration/Appl_Titration2.html [Accessed: 03 March 2017]

[21] Dean GJA. Lange's Handbook of Chemistry. Chapter 8. New York: McGraw-Hill; 1999. pp. 8–113

[22] Rodríguez-Laguna N. Contribuciones teóricas al concepto de capacidad buffer considerando especies insolubles y especies polinucleares en el Sistema [thesis]. Iztapalapa, México, DF: Universidad Autónoma Metropolitana; 2015

[23] Barriada JL, Brandariz I, Sastre de Vicente ME. Acid-base equilibria of monocarboxylic acids in various saline media: Analysis of data using Pitzer equations. Journal of Chemical and Engineering. 2000;**45**:1173–1178

[24] Turner DR, Correia-dosSantos MM, Coutinho P, Gonçalves ML, Knox S. Rapid pK measurements for multibasic weak acids by gradient flow injection titration. Analytica Chimica Acta. 1992;**258**:259–267

[25] Saha A, Saha N, Liang-Lian J, Zhao J, Fridrich G, Sajadi SAA, Song B, Sigel H. Stability of metal ion complexes formed with methyl phosphate and hydrogen phosphate. Journal of Biological Inorganic Chemistry. 1996;**1**:231–238

[26] Martínez-Calatayud J, Campins-Falcó P, Micó-Albert R. Flow method for the titration of weak acids or weak bases using linear titration plots. Analyst. 1987;**112**:1063–1066

[27] Botello JC, Morales-Domínguez E, Domínguez JM, Gutiérrez A, Rojas-Hernández A, Ramírez-Silva MT. A new nuclear magnetic resonance algorithm to determine equilibrium constants of the species in the B(III)-H_2O system. Spectrochimica Acta Part A. 2003;**59**:1477–1486

[28] Matczak-Jon E, Kurzak B, Sawka-Dobrowolska W, Kafarski P, Lejczak B. Zinc(II) complexes of phosphonic acid analogues of glutamic acid. Journal of Chemical Society, Dalton Transactions. 1996;3455–3463

[29] Anderegg G, Sherif K. Extrapolation of molar equilibrium constant to zero ionic strength and parameters dependent on it. Copper(II), nickel(II), hydrogen(I) complexes with glycinate ion and calcium(II), hydrogen(I) complexes with nitrilotriacetate ion. Talanta. 1995;**42**:1067–1079

[30] Christian GD. Analytical Chemistry. New York: Wiley; 2004. p. 237

[31] Ringbom A. Complexation in Analytical Chemistry. New York: Interscience; 1963. p. 395

[32] Rodríguez-Laguna N, Rojas-Hernández A, Ramírez-Silva MT. Study and behavior of the buffer capacity of mixtures of species of a same protons poly-donor system. Educación Química. 2014;**25**(E1):210–222

Chapter 2

The Kjeldahl Titrimetric Finish: On the Ammonia Titration Trapping in Boric Acid

Julia Martín, Lucía Fernández Sarria and
Agustín G. Asuero

Additional information is available at the end of the chapter

http://dx.doi.org/10.5772/intechopen.68826

Abstract

Kjeldahl method using concentrated boric acid is a common practice in many laboratories. A thorough study of the titration with hydrochloric acid of ammonia trapped in a solution of boric acid is made in an attempt to explain the fundamentals of a widely applied standard method. A new potentiometric method for the determination of the end point in the Kjeldahl titrimetric finish is proposed based on the linearization of the titration curve of the ammonia-boric acid system. The method is strictly based on mole and charge balances, and no approximations are made in deriving the equations. The proposed method has proved very accurate when applied to synthetic titration curves and data. Some problems, however, are experienced in the practice, because the behavior of the experimental system studied is far from the expected one on the basis of the theoretical model. However, a slight modification of the devised method has been applied to the experimental titration of ammonia with hydrochloric acid, in boric acid as trapping solution, to get good results.

Keywords: Kjeldahl method, titrimetric finish, ammonia-boric acid system

1. Introduction

Nearly 130 years ago, on March 7, 1883, at a meeting of the Danish Chemical Society, Johan Gustav Christoffer Thorsager Kjeldahl (Head of Chemistry Department of the Carlsberg Foundation Laboratory of the Danish Brewing Carlsberg Company) introduced a method known later under the eponym "Kjeldahl Method" that basically is still in use. Within the same year, the method was published in a German journal Zeitschrift für Analytische Chemie [1] and written in French and Danish languages in communications from the Carlsberg Laboratory.

The Danish brewer Mr. J.C. Jacobsen had a great respect for Pasteur and his work for the French wine industry, and for this reason, French extensive summaries of the Carlsberg papers were also published. As an extended summary of the Kjeldahl paper appeared in Chemical News in August [2], the method was quickly taken up [3]. The Analyst first gave details of the method in 1885 [4, 5], although the method had been briefly mentioned by Blyth [6] though giving the Kjeldahl name incorrectly as Vijeldahl. A surprisingly short period went by between the publication of the Kjeldahl method and the appearance of new publications concerning the topic, both in Europe and America. None of the analytical methods has been as widely chosen, in so short a time [7], as the "Kjeldahl Method" for the estimation of nitrogen.

The Kjeldahl method was originally designed for the brewing industry as an aid in following protein changes in grain during germination. It was Berzelius, who suggested the use of the word "protein" in 1838 in a letter to Mulder because it was derived from the Greek word meaning "to be in the first place" [8]. The Kjeldahl method lacks analytical selectivity because it does not distinguish between protein-based nitrogen from non-protein nitrogen (NPN). Adulteration incidents (i.e., adulteration of protein-based foods with melanine and related nonprotein compounds) exploiting this analytical vulnerability have been recently detected and are new examples of a problem that dates back to before the Kjeldahl method was introduced [9]. The presence of NPN compounds in foods (amino acids, ammonia, urea, trimethylamine oxide) overestimates their true protein content as derived from the current nitrogen determination methods. Separation of NPN from true protein nitrogen may be carried out by addition of a protein precipitating agent such as trichloroacetic acid or perchloric acid.

The protein content in a foodstuff is estimated by multiplying the nitrogen content by a nitrogen-to-protein conversion factor, usually set at 6.25, which assumes the nitrogen content of proteins to be 16%. However, pure proteins differ in terms of their nitrogen contents because of differences in their amino acids composition, ranging from 13.4 to 19.3%. So, different multiplying factors are suitable for samples of different kinds.

The protein content determines the market value of major agricultural commodities. In addition, the quantitative analysis of protein content is necessary for quality control, being also a prerequisite for accurate food labeling. In recent years, soy protein products have increased the interest of consumers, especially in Western cultures, due to the high-quality protein of soy foods together with their associated health benefits [10] (according to FDA [11], 25g of soy protein per day may improve cardiovascular health).

Table 1 [12–32] gives the chemical methods, most commonly used for protein determination. Some of the most significant methods (Dumas, Kjeldahl, and biuret assay) date from the late 1800s. Since the nineteenth century, many other analytical methodologies have been developed to determine the total protein in the field of biochemistry, biology and proteomics, but most of them to address research needs and not necessarily to determine the purity and/or adulteration of food products.

Though there are several experimental approaches to evaluate the nitrogen content in different kinds of samples (Dumas combustion method, NIR methods), the Kjeldahl method still remains as the reference method, being really the "golden standard" for validating other quantifying methodologies in the biopharmaceutical and food industries [9, 33]. The Kjeldahl method is

http://dx.doi.org/10.5772/intechopen.68826

Technique/eponym	References
Dumas method	[13]
Nessler reagent	[14, 15]
Biuret method	[16–18]
Berthelot's method (Alkali-phenol reagent)	[19, 20]
Kjeldahl method	[1, 2]
Folin-Ciocalteu	[21]
Dye binding	[22]
Lowry method	[23–26]
Direct alkaline distillation	[27]
NIR (Near-infrared reflectance)	[27]
Modified Berthelot reaction	[20]
Modified Lowry method	[25, 28]
Bradford method (Coomassie blue dye-binding method)	[29, 30]
BCA (Bicinchoninic acid method)	[31]
3-(4-carboxy benzyl)quinoline-2-carboxaldehyde	[32]

Table 1. Methods for food protein analysis: some key references [12].

applied in official methods [34] to determine the nitrogen content measurement in foods as well as in many other samples, pharmaceutical, agricultural, food products, biological sediments and surface and waste waters. The diversity of papers dealing with Kjeldahl method [12] is attributable to the immense usefulness of the method, to its need for modifications for applications to various types of organic and inorganic compounds, and to the search for catalysts to provide such modifications and to accelerate the digestion [35, 36].

W. Johannsen (1857–1927), a pharmacist, wrote the Kjeldahl obituary, first published in German [37] and then translated to French [38] and English [39]. Johannsen, one of the founders of the science of genetics, was in his beginnings an assistant in the chemistry department at the Carlsberg Laboratory under the chemist Johan Kjeldahl and is well known for coining the term gene in 1909. Kjeldahl was elected to membership in the scientific academies of Denmark and Christiania and received an honorary doctorate from the University of Copenhagen. Kjeldahl (like Nessler) has been verbalized, an honor not usually accorded to a chemist [40]. All chemists understand what it means when it is said that a substance was kjeldahled [5] or that one kjeldahlizes a sample [41]. Kjeldahl was predecessor to S.P.L. Sörensen as Head of the Carlsberg Laboratory in Copenhagen, who introduced the notation of pH [41, 42].

2. The Kjeldahl's three steps

Titration analysis is one of the oldest analytical methods, and as a matter of fact, it plays an important role [43–46] in various analytical fields as well as in routine analysis [47–50]. Generally, the quantity of tested components of a sample is determined by adding to the measured sample an exactly known quantity of the standard titration solution with which the desired constituent reacts in a definite, known proportion. The content of the components is calculated according [45, 46] to the concentration of the standard solution, the consumed volume, the measuring relationship between chemical reactions and the weight of the tested substances, etc. If the tested substance cannot chemically react with the titrant directly, indirect ways of other reactions can be applied to measure its content.

Although some analytical techniques allow the direct determination of species without sample treatment, it is usually necessary to dissolve the sample prior analysis. The analytical adage: "*once the problem is dissolved, the problem is solved*" denotes clearly this fact. Most analyses are preferentially performed on solution samples, a more homogeneous analysis sample representative of the bulk properties of a large solid sample being thus obtained, improving precision and accuracy [51]. Digestion may be defined as the process, in which a complex substance is decomposed into volatile gases and simple salts that are soluble in dilute acid solution [52]. Wet decomposition or acid digestion involves the use of mineral acid, alone or in combination with other acids and oxidizing agents to affect the dissolution of a sample [53].

The three steps [1, 54] of Kjeldahl method includes:

1. Wet acidic (conc. H_2SO_4) digestion (mineralization) of nitrogen-containing sample, in a long-necked digestion flask, causing its conversion into NH_4^+ ions. The Kjeldahl flask, which he constructed in 1888 to simplify the method, is still in use today. Traditionally, Kjeldahl flasks with a capacity for 500–800 mL and gas or electric heating have been used for the digestion. The digest must contain residual H_2SO_4 to retain the NH_3 as NH_4^+. Water is added manually or automatically to prevent the digest from solidifying, which also may cause bumping, and to avoid mixing concentrated alkali with concentrated acid [55] during the distillation.

2. Transformation of NH_4^+ ions into NH_3(neutralization with alkali) and distillation of the NH_3. The flask is heated after the addition of water and alkali to the digested sample, in order to distill a volume of distillate and collect NH_3 in the acidic distillation receiver. Direct steam distillation drastically decreases the time needed for distillation [55]. Experimental conditions including rate of steam flow, vigor of distillation and volume of solution must be correctly balanced to ensure efficient removal of ammonia without any carry-over of alkaline spray.

3. Titration of the solution from the distillation receiver. The ammonia from the distillation is frequently collected in an excess of standard acid and determined by a back titration with standard alkali solution. A more common practice is the use of boric acid for trapping ammonia. The titration in this case should be carried out as soon as possible after the distillation is complete, ensuring that the temperature of the distillate does not exceed 25°C. Under these conditions, losses of ammonia are avoided [56] (EN ISO 5983-1, 2009).

Complete details [12] of the three Kjeldahl steps are too lengthy to include here. A series of factors such as sample origin, homogeneity, stability, laboratory skillfulness, sample handling procedures, and composition are of critical importance, as are the size of the test portion taken for analysis, and concentration of the titrant used in the Kjeldahl analysis [57]. A relatively large analytical sample (1–2g) was used in the original method, requiring large amounts of acid.

Green alternatives exert considerable pressure to ensure the safe disposal of mercury (when used as catalyst) and, especially, to minimize acid usage [58–60]. The use of Micro-Kjeldahl methods is common practice in order to reduce the amount of acid fumes and also require less acid and catalyst mixture.

The introduction of aluminum blocks (at Tecator, now FOSS) in the early 1970s [61] made possible to improve the speed and accuracy of the digestion procedure, thus saving space, chemicals and energy [60]. The digestion system has since been improved, and the distillation step has been speeded up by the use of the semi-automated systems now available. Block digestions followed by steam distillation are named as rapid Kjeldahl. As a result of technical innovations, there are also available fully automated protein analysis systems that are based on the classical Kjeldahl procedure, for example, the Kjeltec series of Foss Tecator manufactures.

3. The ammonia determination: titrimetric finish

Transformation of NH_4^+ into NH_3 caused by addition of NaOH (pH growth) into Pregl-Parnas-Wagner results [12] from equation

$$\frac{[NH_3]}{[NH_4^+]} = 10^{pH-pK_{1,a}} \tag{1}$$

where $pK_{1,a} = -\log K_{1,a}$; $K_{1,a}$ refers to reaction $NH_4^+ = H^+ + NH_3$, $pK_{1,a} = 9.35$ at 20°C.

The distillation titration method is a standard procedure used by most laboratories to measure ammonium nitrogen in the total Kjeldahl nitrogen digests of various kinds of agricultural and environmental samples [62]. Ammonia may be collected into a solution of a strong acid (HCl or H_2SO_4). After distillation, the excess of standard acid may be iodometrically determined with starch as indicator, as was done by Kjeldahl, according to consecutive reactions $IO_3^- + 5I^- + 6H^+ = 3\,I_2 + 3H_2O$; $I_2 + 2\,S_2O_3^{2-} = S_4O_6^{2-} + 2\,I^-$, though this method is seldom used. The excess of standard acid is instead usually titrated with a strong base. Two standard solutions are then needed: (1) titrant (NaOH solution) and (2) the strong acid in the receiver.

Titration of ammonia absorbed in H_3BO_3 solution was proposed first by Winkler [63] which notes "*Boric acid is indeed such a week acid, that its solution does not cause a noticeable colour change of certain indicators. Ammonia is, however, completely fixed by it, provided that a suitable excess of acid is present.*" Boric acid is also commonly used to trap ammonia because only one standardized solution (e.g., HCl or H_2SO_4) is needed (as titrant).

Mixtures of methyl red and tetrabromophenol blue or methyl red and methylene blue (2:1) (Tashiro indicator) or methyl red alone were suggested as indicators; the latter changes its color at about *pH* 5.2. After addition of HCl, the Tashiro indicator changes its color from green to violet. Alternatively, a *pH* titration with end point (V_{eq}) at a preset *pH* value of 5.0 is also done. A mixed indicator (bromocresol green and methyl red) has been recently used by Beljkaš et al. [64] to locate the end point.

4. Base or acid as titrant: the boric acid trapping choice

The ammonia can be distilled into an excess of standard strong acid (HCl or H_2SO_4), and the excess determined by back titration with a strong standard base, since the solution at the equivalence point contains $(NH_4)_2SO_4$, or NH_4Cl, which hydrolyses,

$$pH = -\log\sqrt{K_a C_A} = \frac{pK_a + pC_A}{2} = \frac{9.36 + 1.30}{2} = 5.33 \tag{2}$$

(i.e., 0.05M of ammonium chloride at the end point), methyl red (transition range 4.4–6.2) being used as an indicator [65].

A mistake appears in the (excellent) Bradstreet monograph [66], p. 152, who wrote "*in the case of back titration of a distillate, the equivalence point will occur at* pH *7, since this is the point of neutralization of a strong acid by a strong base. Any indicator, therefore, changing colour at or close to* pH *7 is suitable.*" The millimoles of NH_3 in the sample are equal to the total millimoles of HCl added minus millimoles of base used in back titration. This method has two disadvantages: i) the amount of NH_3 is obtained as a difference; and ii) two standard solutions are required.

Preferably, the distilled ammonia can be absorbed into a solution of boric acid [63] or other weak acid [67]. Boric acid is sufficiently acid to react with ammonia and prevent loss by volatilization, but it is apparently too weak an acid to interfere with the titration of ammonium borate with diluted hydrochloric acid. When boric acid is used instead of standard acid as the recipient of the distillate, the use of a weak standard alkaline solution is avoided, which suppose a distinct advantage. The alkaline solution is vulnerable to the absorption of atmospheric carbon dioxide with subsequent changes in molarity. Neither the amount nor the concentration (about 4%) of boric acid in the receiving bottle has to be precise.

Then, in the Winkler modification, the NH_3 is caught in an unmeasured excess of boric acid

$$NH_3 + H_3BO_3 = NH_4^+ + H_2BO_3^- \tag{3}$$

The borate formed is determined by titration with standard HCl, one mole of HCl being required for each mole of NH_3

$$H_2BO_3^- + H^+ = H_3BO_3 \tag{4}$$

The solution at the equivalence point contains H_3BO_3 and NH_4Cl, a mixture of two weak acids,

$$pH = -\log\left(\sqrt{K_{a1}C_{A1} + K_{a2}C_{A2}}\right) \tag{5}$$

so that an indicator transiting in the acid region (*pH* 5–6) is satisfactory [65].

Note that the acidity constants of boric acid and ammonium ion are very similar. The equilibrium constant of Eq. (3) is given by

$$K_{eq} = \frac{[NH_4^+][^-H_2BO_3]}{[NH_3][H_3BO_3]} = \frac{K_{1,H_3BO_3}}{K_{1,NH_4^+}} = \frac{10^{-9.24}}{10^{-9.35}} = 1.288 \tag{6}$$

Though this equilibrium constant is low, the fraction of ammonia converted into ammonium ion increases with increasing the difference of concentrations between boric acid and ammonia. As we are studying the trapping process (without titrant added), the volume may take as constant and the mass balances are given by

$$C_{H3BO3} = [H_3BO_3] + [H_2BO_3^-] \tag{7}$$

$$C_{NH3} = [NH_3] + [NH_4^+] \tag{8}$$

The following relationship is satisfied in the trapping solution (if solution is not very diluted, the contribution of dissociation of water being in those cases is negligible).

$$[NH_4^+] = [H_2BO_3^-] \tag{9}$$

and then, by combining Eqs. (6) to (9), we get

$$K_{eq} = \frac{[NH_4^+]^2}{(C_{NH_3} - [NH_4^+])(C_{H_3BO_3} - [NH_4^+])} \tag{10}$$

which on rearrangement gives the second degree equation

$$\left(1 - \frac{1}{K_{eq}}\right)[NH_4^+]^2 - (C_{NH_3} + C_{H_3BO_3})[NH_4^+] + C_{NH_3}C_{H_3BO_3} = 0 \tag{11}$$

which may be solved for given concentrations of boric acid and ammonia. Once the value of $[NH_4^+]$ is known, that is, the fraction of ammonium ion

$$f_{1,\alpha} = \frac{[NH_4^+]}{C_{NH_3}} \tag{12}$$

the *pH* values may be calculated from Eqs. (2) and (12)

$$[H^+] = K_{1,\alpha}\frac{[NH_4^+]}{[NH_3]} = K_{1,\alpha}\frac{f_{1,\alpha}}{f_{0,\alpha}} = K_{1,\alpha}\frac{f_{1,\alpha}}{1 - f_{1,\alpha}} \tag{13}$$

Hydrochloric acid titrates then the borate ion from the above reaction (Eq. (4)) as well as the ammonia, which is not converted into ammonium ion in Eq. (3), that is, $H_2BO_3^- + H^+ = H_3BO_3$; $NH_3 + H^+ = NH_4^+$ (the sum of both being equivalent to the ammonia distilled).

The trapping process may be illustrated by means of semilogarithmic diagrams (**Figure 1**). **Figure 1** shows the fraction of ammonium ion (equivalent to the borate ion), at a given $p[NH_3]$ for varying concentrations of trapping boric acid. In the absence of other equilibria than those considered previously, this diagram would help to know the concentration of ammonia from the pH initial of the sample to be titrated, that is, a pH of about 7 for a 2% boric acid trapping solutions corresponds to a $p[NH_3]$ of about 2. However, in numerous cases, things are not an easy business as may appear at first sight and apparently some complications arise in the boric acid-ammonia system associated when high boric acid concentrations are present. When significant quantities of ammonia vapor are delivered into 4% aqueous boric acid containing methyl red-methylene blue indicator, it is necessary to dilute with distilled water so that the expected change in color (purple to green) may occur. Only after considerable dilution had been effected, could the ammonia [68] be satisfactorily titrated with hydrochloric acid. Boric acid forms [68, 69] polymeric borate species (triborate, tetraborate, pentaborate) in concentrated solutions whose acidity constants are greater than the $K_{1,b}$ of boric acid by a factor between 210 and 440, thus boric acid behaving as a stronger acid in concentrated solution than in diluted solution.

As far as we know, the anomaly behavior of boric acid was revealed first by Prideaux [70] one century ago when studying the titration curve of boric acid, and then by Thygesen [71] through conductivity measurements. In spite of the fact that Kjeldahl titration is one of the titrimetric methods more applied worldwide, scarce mention of this fact has been made in the literature. However, some problems posses the location of the end point in Kjeldahl titration

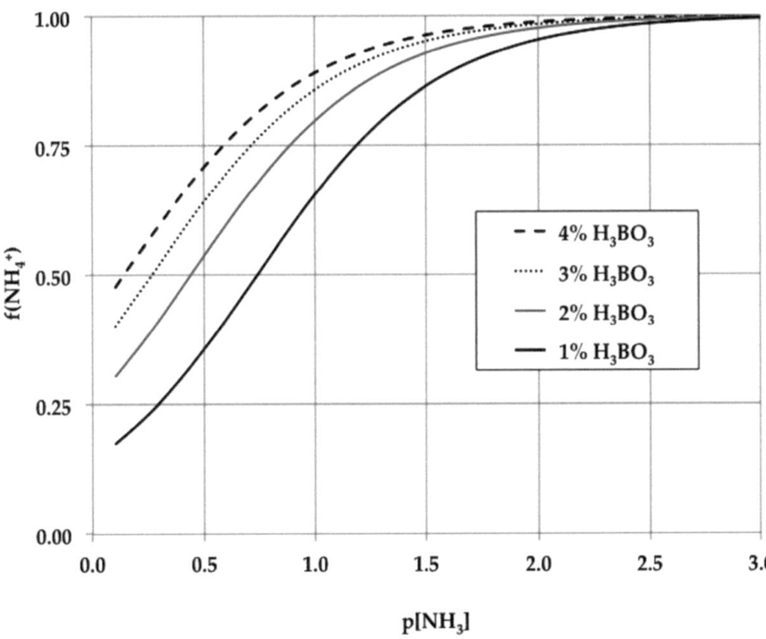

Figure 1. Semilogarithmic diagram molar fraction of ammonium ion versus minus logarithm of the total ammonia concentration ($p[NH_3]$), at varying percentage of boric acid as trapping solution.

http://dx.doi.org/10.5772/intechopen.68826

for the indicators recommended, for example, bromophenol blue, methyl red, methyl orange, and congo red, as we have previously indicated.

This fact is another argument in favor to carry out a potentiometric study of the ammonia-boric acid system with the aim of devising a straight-line linearization of titration curve in order to accurately locate the end point of the titration curve. The theoretical background for the new method is outlined in the following sections.

5. Titration curves in the ammonia-boric hydrochloric acid titration system

A thorough study of the titration of ammonia trapped in a solution of boric acid is made in this section based on a mathematical approach with equations derived for the titration curves involved. The calculations presented are based on charge and concentration balances, and expressions for equilibrium constants related to acid-base equilibria. This approach can be perceived as the clear confirmation of the statement [72], ascribed to J.C. Maxwell that *"a good theory is the best practical tool."*

Let us consider the titration of ammonia trapped in boric acid solution with hydrochloric acid as titrant. The initial concentration of ammonia and boric acid is denoted by C_{NH3} and C_{H3BO3}, respectively, and the initial volume by V_0, C_{HCl} denotes the concentration of strong acid and V the volume added. The electroneutrality rule for the solution is

$$[H^+] + [NH_4^+] = [OH^-] + [Cl^-] + [^-H_2BO_3] \tag{14}$$

Eq. (14) should also include concentrations of ions being originated from the neutral salt added to adjust ionic strength (when necessary). As the salts used for this purpose usually potassium nitrate and chloride are completely dissociated, the concentration of the anion and cation is the same and cancel out in the equation. Also, the contribution of the second and third dissociation steps of boric acid is considered negligible at the working pH range.

The law of mass action holds for the reactions

$$NH_4^+ \rightleftharpoons NH_3 + H^+ \qquad K_{1,a}^c = \frac{[NH_3][H^+]}{[NH_4^+]} \tag{15}$$

$$H_3BO_3 \rightleftharpoons {}^-H_2BO_3 + H^+ \qquad K_{1,b}^c = \frac{[^-H_2BO_3][H^+]}{[H_3BO_3]} \tag{16}$$

The constants K_{1i} $(i=a, b)$ are concentration constants, useful when the pH-meter is calibrated in term of hydrogen ion concentrations at an ionic strength fixed. If this not the case, mixed constants (at an ionic strength fixed), where the activity of the hydrogen and hydroxide ions (denoted by (H^+) and (OH^-)) are used in conjunction with concentrations of all other species. The activity of the hydrogen ion is related to concentration by the expression

$$(H^+) = [H^+]\gamma_H \tag{17}$$

where γ_H is the activity coefficient of the ion. Then

$$K^B_{1,a} = \frac{[NH_3](H^+)}{[NH_4^+]} = K^C_{1,a}\,\gamma_{H^+} \qquad K^B_{1,b} = \frac{[^-H_2BO_3](H^+)}{[H_3BO_3]} = K^C_{1,b}\,\gamma_{H^+} \tag{18}$$

The total concentrations of the ammonia and boric acid during the titration can be expressed by their respective mass balances

$$[H_3BO_3] + [^-H_2BO_3] = \frac{V_0 \cdot C_{H_3BO_3}}{V_0 + V} \tag{19}$$

$$[NH_4^+] + [NH_3] = \frac{V_0 C_{NH_3}}{V_0 + V} \tag{20}$$

The concentration of the chloride ion can be expressed by the following equation

$$[Cl^-] = \frac{V \cdot C_{HCl}}{V_0 + V} \tag{21}$$

The molar fractions of ammonia and borate ions are given by

$$f_{1,a} = \frac{[NH_4^+]}{[NH_3] + [NH_4^+]} = \frac{\frac{[NH_4^+]}{[NH_3]}}{1 + \frac{[NH_4^+]}{[NH_3]}} = \frac{\frac{[H^+]}{K^c_{1,a}}}{1 + \frac{[H^+]}{K^c_{1,a}}} = \frac{10^{pK^c_{1,a}-pH}}{1 + 10^{pK^c_{1,a}-pH}} \tag{22}$$

$$f_{0,b} = \frac{[^-H_2BO_3]}{[H_3BO_3] + [^-H_2BO_3]} = \frac{1}{1 + \frac{[H_3BO_3]}{[^-H_2BO_3]}} = \frac{1}{1 + \frac{[H^+]}{K^c_{1,b}}} = \frac{1}{1 + 10^{pK^c_{1,b}-pH}} \tag{23}$$

and the ammonia and borate concentrations expressed in terms of their respective molar fractions may be substituted into Eq. (14), giving

$$[H^+] + f_{1,a}\frac{V_0 \cdot C_{NH_3}}{V_0 + V} = [OH^-] + \frac{V \cdot C_{HCl}}{V_0 + V} + f_{o,b}\frac{V_0 \cdot C_{H_3BO_3}}{V_0 + V} \tag{24}$$

and then we get

$$\Delta = [H^+] - [OH^-] = \frac{V \cdot C_{HCl} + f_{o,b} V_0 \cdot C_{H_3BO_3} - f_{1,a} V_0 \cdot C_{NH_3}}{V_0 + V} \tag{25}$$

From the Eq. (25), we can, after a simple mathematical manipulation, obtain

$$V = V_0 \frac{f_{1,a} C_{NH_3} - f_{o,b} C_{H_3BO_3} + \Delta}{C_{HCl} - \Delta} \tag{26}$$

thus obtaining a relatively simple, closed-form expression for the titration curve valid at any moment of the titration. It relates the volume of added titrant to the hydrogen ion concentration (*p[H]*) of the solution. Then, by substituting a series of values of *[H⁺]*, we calculate the corresponding values of *V*.

6. Graphical method for the determination of the equivalence point

Numerical methods based on the mathematical modeling of the titration curve may be applied for the determination of the end point instead of approximate methods. Gran's linearization known since 1950 [73–75] is one of the examples. In this book chapter, a new method for the determination of the end point is proposed based on the linearization of the titration curve of the ammonia-boric acid system. The method is inspired in a previous method described first by Schwartz [76] and also explained with great detail in two relatively recent analytical chemistry textbooks [77, 78].

Eq. (25) gives

$$(V_0 + V)\Delta = V \cdot C_{HCl} + f_{0,b} V_0\, C_{H_3BO_3} - f_{1,a} V_0 C_{NH_3} \tag{27}$$

and dividing the left and right members through by C_{HCl} we have

$$\frac{(V_0 + V)\Delta}{C_{HCl}} = V + \frac{f_{0,b} V_0 \cdot C_{H_3BO_3}}{C_{HCl}} - f_{1,a} V_{eq} \tag{28}$$

The following equation should be valid at the equivalence point

$$V_0\, C_{NH_3} = V_{eq} C_{HCl} \tag{29}$$

and then, by combining Eqs. (31) and (30), we get

$$V - \frac{(V_0 + V)\Delta - f_{0,b} V_0 \cdot C_{H_3BO_3}}{C_{HCl}} = f_{1,a} V_{eq} \tag{30}$$

The left member of Eq. (32) is denoted by V'

$$V' = V - \frac{(V_0 + V)\Delta - f_{0,b} V_0 \cdot C_{H_3BO_3}}{C_{HCl}} \tag{31}$$

and taking into account that Eq. (22) is equivalent to

$$f_{1,a} = \frac{[H^+]}{[H^+] + K_{1,a}} \tag{32}$$

we get

$$V' = \frac{[H^+]}{[H^+] + K_{1,a}} \tag{33}$$

By simple manipulation, Eq. (35) gives

$$V'([H^+] + K_{1,a}) = [H^+]V_{eq} \quad V'K_{1,a} = [H^+]V_{eq} - V'[H^+] \tag{34}$$

By multiplying the left and right members of Eq. (36) through by $1/(K_1,a\ [H^+])$ we finally have

$$\frac{V'}{[H^+]} = \frac{V_{eq}}{K_{1,a}} - \frac{V'}{K_{1,a}} \tag{35}$$

Eq. (35) gives a straight line when $V'/[H]$ is plotted against V'. The plot has a slope of $-1/K_{1,a}$ and intersects the V' axis at the point V_{eq}. By use of Eq. (35) the equivalence point can be located with considerable accuracy. Eq. (35) may be considered as a variant (an extension) of the method of Schwartz [76] for the determination of the equivalence point in the titration of a weak base with a strong acid.

The method, which we propose, is based strictly on mole and charge balance equations. No approximations are made in deriving Eq. (35). Both the dissociation of water as well as the dilution effects is precisely accounted for. The only restrictions of the method are, on the one hand, the accuracy, with which the pH-meter and electrodes can be calibrated, and several fundamental assumptions entailed by linear regression that are not always satisfied with data obtained. One, for example, is that the values of x are free from error but those of y are drawn from a population having normally distributed errors [79, 80]. When the precision of experimental measurements is very high, no special problems, however, are to be expected.

Note that the use of the complete Eq. (31), which gives the modified volume function V', presupposes knowledge of the $pK_{1,b}$ necessary to obtain the $f_{0,b}$ value (Eq. (23)). In any case, the $pK_{1,b}$ value may be extracted from appropriate tables, or from the analytical bibliography [81], though it is not necessary to be known exactly. A trial value of $pK_{1,b}$ may be assumed and values for V_{eq} and $K_{1,a}$ calculated by least squares method on Eq. (35). The procedure is repeated for other assumed values of $pK_{1,b}$, and the best value is taken as that for which minimizes $s_{y/x}$ the standard deviation of the corresponding regression line. The $s_{y/x}$ values can be easily got using linear regression (method of the least squares), in EXCEL, with the function LINEST.

Taking into account Eqs. (3) and (4), the computation of ionic strength may be made as a first approximation, when $V < V_{eq}$, as

$$I = \frac{1}{2}\left(\frac{2C_{NH_3}V_0 - C_{HCl}V}{V_0 + V} + [OH^-]\right) \tag{36}$$

and when $V > V_{eq}$ as

$$I = \frac{1}{2}\left(\frac{C_{NH_3}V_0}{V_0 + V} + [H^+]\right) \tag{37}$$

As the activity coefficient γ_H will change only to a very small extension during a titration, log γ_H may be considered constant. A large ratio of sample volume to titrant volume (i.e., 10:1) minimizes errors introduced by variations in activity coefficients due to dilution. The value of most constants has been determined at an ionic strength of I=0.1. It is thus practical to adjust the ionic strength to this value. However, addition of a basal electrolyte into the solution and the titrant to keep the ionic strength approximately constant and rather high is not usually practiced [82] in the titrations involved with Kjeldahl method.

$\gamma_H \approx 1$ is assumed for simplicity of considerations when the titrations are carried out at low ionic strength. The success of the system of pH standardization depends on the validity of putting–log (H^+) equal to pH. As $[H^+]$ normally makes only a small contribution, no sensible error is introduced if (H^+) is used in its place: in some exceptional cases, when $[H^+]$ is relatively more important, the quantity may be estimated with sufficient exactness from Eq. (17); on the acid side of pH (when $[OH^-]$ is negligible).

If we represent the pH as $A \pm \varepsilon$, where ε is the uncertainty, then we have $[H^+]=10^{-(A\pm\varepsilon)}=10^{-A}$ times $10^{\pm\varepsilon}$, where $10^{\pm\varepsilon}=1$ for $\varepsilon=0$ [77]. The resulting uncertainty in $[H^+]$ is now a multiplicative factor, and therefore a relative uncertainty, which applies regardless of the value of pH. A relatively small error in the pH of 0.01, 0.02, or 0.05 corresponds to an uncertainty in $[H^+]$ of about 2, 5 or 12%, respectively. Consequently, Eq. (35) plot is characterized by extreme sensitivity to small changes in pH. This problem is compounded by possible activity effects and the requirement to use a precise value of $pK_{1,b}$.

Differential (approximate) methods for the determination of end points of titration are based on the presumption that the end point of a titration is the inflection point of the titration curve [83, 84], where the absolute value of the first derivative reaches a maximum (titration of a weak acid with a strong base) or a minimum (titration of a weak base with a strong acid) and the second derivative changes sign. A number of points very closely spaced and preferably of high precision are needed [83, 84] in order that the method is successfully applied. Only the points in the vicinity of the inflection are used for the calculation. This may result in increasing errors as titration data are least accurate right near the end point [85], because buffering is minimal and electrode response is sluggish. The local pH fluctuations in the inflection region are mostly due [77] to insufficiently rapid mixing of the titrant and the sample, and localized pH sampling by the glass electrode.

Two advantages may be ascribed to the method devised in this book chapter, based on the Eq. (35). Fewer titration points need to be taken than with conventional methods, and measurements need not be made close to the equivalence point since this point may be obtained by extra(inter)polation. Therefore, problems related to incompleteness of reaction or instability of measurements close to the end point might be avoided. Note that when very weak acids or bases are titrated, the approximations assumed by Gran when deriving their equations are no longer valid.

Harris [85] have comment in reference to the linearization of titration curve by Gran [74]: "*The beauty of a Gran plot is that it enables us to use data taken before the end point to find the end point.*" This sentence is undoubtedly applied to linear extrapolation methods based [76–78, 86] on an improvement in the methodology proposed by Gran, as it is the case in this book chapter.

7. Approximation expressions derived from the complete equation

Eq. (31) may be expressed as

$$V' = V - \frac{(V_0 + V)\Delta}{C_{HCl}} + \frac{f_{0,b} V_0 \cdot C_{H_3BO_3}}{C_{HCl}} \tag{38}$$

Which is in the form

$$V' = V + U + W \tag{39}$$

$$U = -\frac{(V_0 + V)\Delta}{C_{HCl}} \tag{40}$$

$$W = \frac{f_{0,b} V_0 C_{H_3BO_3}}{C_{HCl}} \tag{41}$$

Figure 2 depicts the contributions of the different terms in Eq. (39) to the value of V'. It is valid for a titration of 150 mL of 0.0096 M ammonia solution and 0.2156 M trapping acid boric solution, titrated with HCl 0.12 M ($pK_{1,a} = 9.27$; $pK_{1,b} = 9.12$; $pK_w{}^c = 13.80$). It is seen that before the equivalence point, the U term may be considered as negligible whereas beyond the equivalence point it is the W term, which may be neglected.

Thus, before the equivalence point we approximately get

$$V' = V + \frac{f_{0,b} V_0 \cdot C_{H_3BO_3}}{C_{HCl}} \tag{42}$$

Note that the mole fraction of borate varies from about 0.043 at the beginning of the titration to about 0 at the equivalence point (**Figure 3**). Despite the fact that initial volume V_0 and concentrations of HCl and H_3BO_3 are constants, the contribution of the second term of the right hand of Eq. (42) is null only when approximated at the equivalence point. There is no way to use a conventional Gran equation method based on the quotient *V*/*[H]* in the ammonia-boric acid titration case.

http://dx.doi.org/10.5772/intechopen.68826

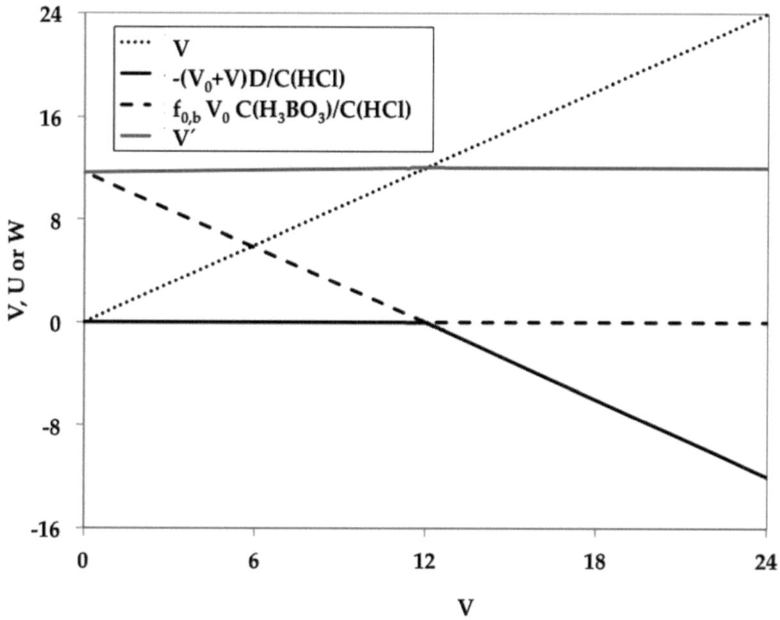

Figure 2. Contributions of the different terms in Eq. (38) to the value of V'.

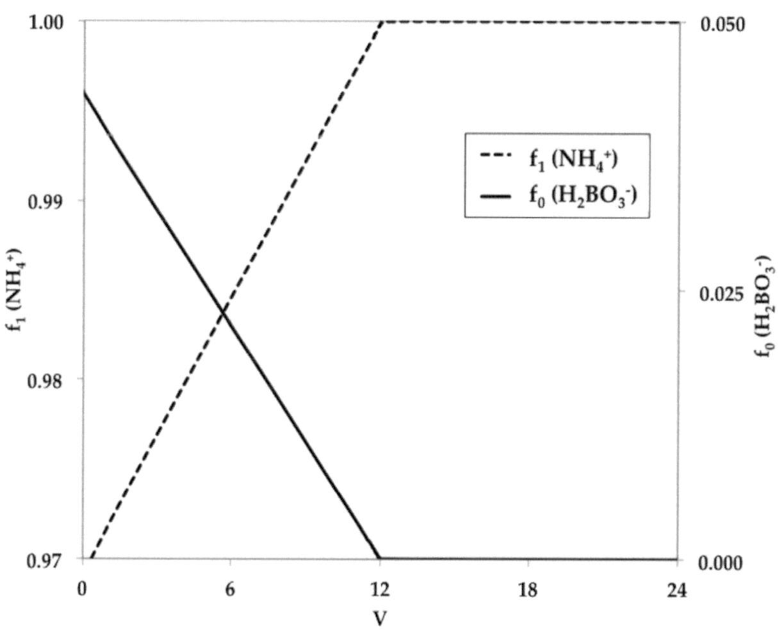

Figure 3. Mole fraction of ammonium ($f_{1,a}$) and borate ($f_{0,b}$) ions as a function of V.

Beyond the equivalence point we have instead

$$V' = V - \frac{(V_0 + V)\Delta}{C_{HCl}} \tag{43}$$

When $V > V_{eq}$, $[OH^-]$ is negligible, and

$$\Delta = [H^+] - [OH^-] \approx [H^+] \tag{44}$$

and $f_{1,a}$, the mole fraction of ammonium ion is virtually equal to the unity, and then by combining Eqs. (43), (44) and (35) we get

$$V' = V - \frac{(V_0 + V)[H^+]}{C_{HCl}} = V_{eq} \tag{45}$$

and then we obtain the Gran expression for the titration of a weak base with a strong acid, beyond the equivalence point

$$(V_0 + V)[H^+] = V\,C_{HCl} - V_{eq}C_{HCl} \tag{46}$$

By plotting the left hand of Eq. (46) against V, a straight line is obtained with slope C_{HCl}, intersecting the V axis at the point V_{eq}.

An alternative route to obtain Eq. (46) is from Eq. (27). At $V > V_{eq}$, $f_{0,b} \approx 0$, $f_{1,a} \approx 1$, $\Delta \approx [H^+]$, and then,

$$(V_0 + V)[H^+] \approx V\,C_{HCl} - V_0 C_{NH_3} \tag{47}$$

which is combined with Eq. (29), gives finally Eq. (46).

8. Titration error

From Eq. (26), we may calculate the fraction titrated, T, as

$$T = \frac{C_{HCl}V}{C_{NH_3}V_0} = \frac{C_{HCl}}{C_{NH_3}}\left(\frac{f_{1,a}C_{NH_3} - f_{0,b}C_{H_3BO_3} + \Delta}{C_{HCl} - \Delta}\right) = \frac{f_{1,a} - \frac{f_{0,b}C_{H_3BO_3}}{C_{NH_3}} + \Delta}{1 - \frac{\Delta}{C_{HCl}}} \tag{48}$$

In the vicinity of the equivalence point, the mole fraction of borate ion, $f_{0,b}$, is close to zero, and the titration error, ΔT, may be approximated as

$$\Delta T = [T - 1]_{end} = \left[\frac{f_{1,a} + \Delta}{1 - \frac{\Delta}{C_{HCl}}} - 1\right]_{end} \tag{49}$$

http://dx.doi.org/10.5772/intechopen.68826

9. Checking the proposed linearization method with synthetic data

In order to verify the goodness of the proposed linearization method, based on the use of Eq. (35), a series of theoretical data with variations of 0.05 units of *pH*, has been generated with aid of Eq. (26), using the analogous conditions previously published by Cruz [68], V_0=150mL; C_{NH3}=0.0096M; C_{H3BO3}=0.2156M; C_{HCl}=0.12M; $pK_{1,a}$=9.27; $pK_{1,b}$= 9.12.

Two sets of data [*V, pH*] (N=24 and N=11) were selected from the theoretical data in the region prior to the equivalence point (V_{eq}=12mL), ranging from approximately 0.6 to 11.16mL for the first series, and from 0.6 to 10.8mL for the second (**Table 2**). Note that the points selected for study are far from the region of the equivalence point: $V_{eq}\pm$1mL. A set of data [*V, pH*] (N=11) was also selected after the equivalence point, between about 13 and 23.5mL (**Table 3**).

The proposed linearization method has been applied to data corresponding to $V < V_{eq}$ (basic zone), the volumes being rounded to 5, 3 and 2 decimal places, respectively. It is proved that the method works well in all cases, correct values for both V_{eq} and $pK_{1,a}$ obtained. The same rounding procedure has been applied to the data corresponding to $V > V_{eq}$ (acid zone).

Applying in this latter case, the Gran method based on Eq. (46) also leads to the correct values of V_{eq} and C_{HCl}. However, a pronounced curvature is observed when the conventional Gran

	V_0=	150 mL			V_{eq}=	12.000 mL	
	C(HCl)=	0.12M			$pK_{1,a}$=	9.270	
	$pKa(NH_4^+)$=	9.27					
	$pKa(H_3BO_3)$=	9.12		a_1=	-1.864E+09	2.237E+10	$=a_0$
	$C(H_3BO_3)$=	0.2156M		$s_{(a1)}$=	1.971E+07	2.327E+08	$=s_{(a0)}$
	pKw=	13.8		R^2=	0.9990	6.386E+06	$=s_{(y/x)}$
V	**V(2)**	**V(2)/[H]**	**pH**	**Δ**	$\mathbf{f_{0,b}}$	**V′(2)**	**v′(2)/[H]**
0.62116	0.62	3.49E+07	7.75	-8.73E-07	4.09E-02	11.65	6.55E+08
1.81368	1.81	9.07E+07	7.70	-7.74E-07	3.66E-02	11.68	5.85E+08
2.88545	2.89	1.29E+08	7.65	-6.86E-07	3.28E-02	11.72	5.24E+08
3.84785	3.85	1.53E+08	7.60	-6.06E-07	2.93E-02	11.75	4.68E+08
4.71136	4.71	1.67E+08	7.55	-5.34E-07	2.62E-02	11.77	4.18E+08
5.48560	5.49	1.74E+08	7.50	-4.70E-07	2.34E-02	11.80	3.73E+08
6.80061	6.80	1.71E+08	7.40	-3.58E-07	1.87E-02	11.84	2.97E+08
7.85420	7.85	1.57E+08	7.30	-2.66E-07	1.49E-02	11.87	2.37E+08
8.69687	8.70	1.38E+08	7.20	-1.88E-07	1.19E-02	11.90	1.89E+08
9.65355	9.65	1.08E+08	7.05	-8.87E-08	8.44E-03	11.92	1.34E+08
10.81924	10.82	6.08E+07	6.75	8.87E-08	4.25E-03	11.96	6.73E+07

Table 2. Proposed method applied to theoretical data before the equivalence point and rounded to the second decimal figure in volume (penultimate and last columns).

V_0=	150 mL			Veq=	11.999 mL	
C(HCl)=	0.12M			C(HCl)=	0.120M	
$pKa(NH_4^+)$=	9.27					
$pKa(H_3BO_3)$=	9.12		a_1=	0.120	−1.440	$=a_0$
$C(H_3BO_3)$=	0.2156M		$s_{(a1)}$=	4.471E-05	8.219E-04	$=s_{(a0)}$
pKw=	13.8		R^2=	0.9999	4.778E-04	$=s_{(y/x)}$
V	**(V_0+V) [H]**	**V(3)**	**$(V(3)+V_0)$ [H]**	**V(2)**	**$(V(2)+V_0)$ [H]**	**pH**
13.07922	0.129538429	13.079	0.130	13.08	0.13	3.10
14.16181	0.260178937	14.162	0.260	14.16	0.26	2.80
15.07963	0.369567257	15.080	0.370	15.08	0.37	2.65
15.89625	0.467559159	15.896	0.468	15.90	0.47	2.55
16.93586	0.592310783	16.936	0.592	16.94	0.59	2.45
18.26332	0.751604642	18.263	0.752	18.26	0.75	2.35
19.06089	0.847311598	19.061	0.847	19.06	0.85	2.30
19.96482	0.955782421	19.965	0.956	19.96	0.96	2.25
20.99062	1.078877875	20.991	1.079	20.99	1.08	2.20
22.15642	1.218774118	22.156	1.219	22.16	1.22	2.15
23.48355	1.37802882	23.484	1.378	23.48	1.38	2.10

Table 3. Gran method applied to theoretical data after the equivalence point and rounded to the second decimal figure in volume (antepenultimate and penultimate columns).

function ($V/[H^+]$) is applied to the data obtained in the basic region ($V < V_{eq}$), which makes it impossible to apply this method to values of the ammonia-boric acid system away from pH_{eq}.

From the data obtained and using linear regression (method of the least squares), in EXCEL, with the function LINEST, the volume at the equivalence point is obtained as

$$V_{eq} = -\frac{a_0}{a_1} \qquad pK_{1,a} = -\log\left(-\frac{1}{a_1}\right) \tag{50}$$

to data corresponding to $V < V_{eq}$, and

$$V_{eq} = -\frac{a_0}{a_1} \qquad C_{HCl} = a_1 \tag{51}$$

to data corresponding to $V > V_{eq}$. The data corresponding to the statistic parameters of the straight line using the least squares method are displayed in **Tables 2** and **3**. The parameters obtained using linear regression by the function LINEST in EXCEL are: intercept, slope and coefficient of determination (correlation coefficient squared): a_0, a_1 and R^2, respectively; and standard deviations of the intercept, slope and regression: $s_{(a0)}$, $s_{(a1)}$ and $s_{(y/x)}$, respectively.

http://dx.doi.org/10.5772/intechopen.68826

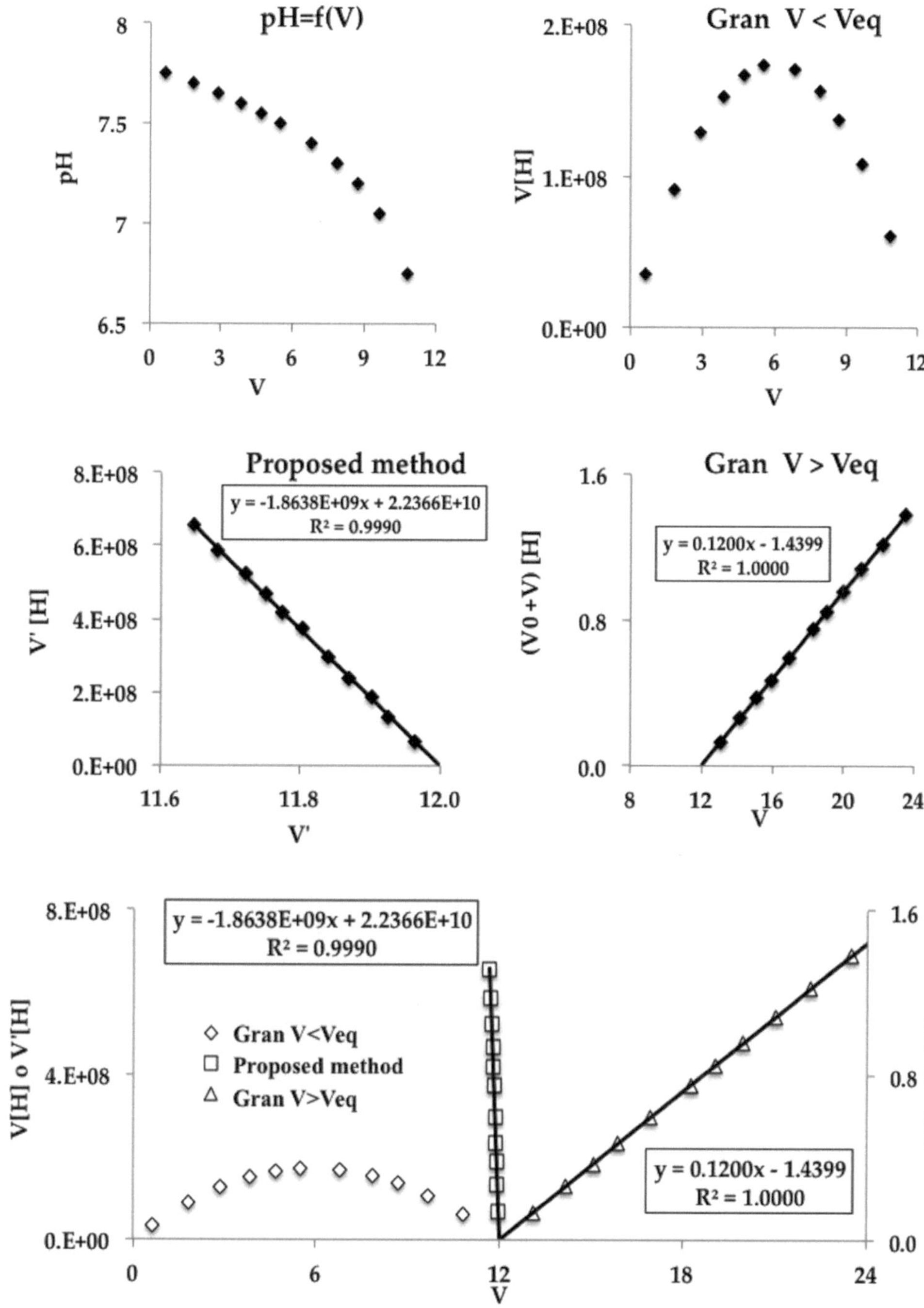

Figure 4. Top: ammonia titration curve ($pH=f(V)$) in boric acid (LEFT) and Gran method $V < V_{eq}$ (right). Middle left: modified Schwartz method. Middle right: Gran method $V > V_{eq}$. Bottom: comparative study.

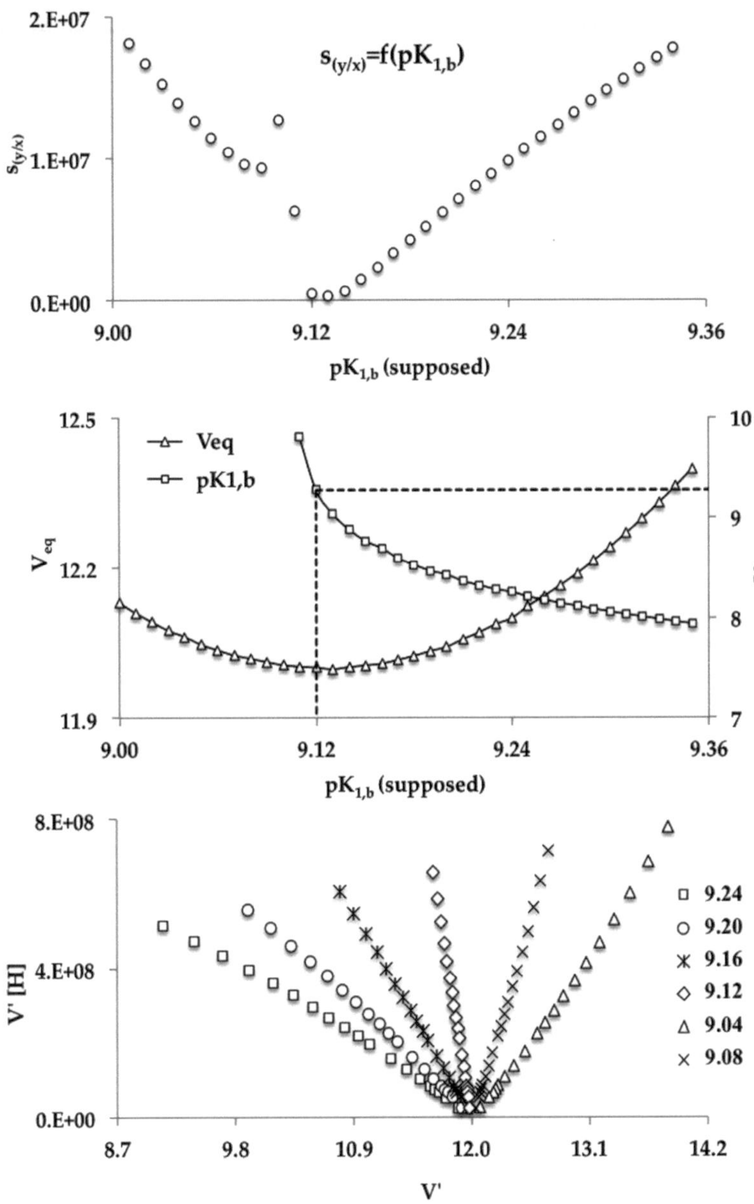

Figure 5. Top: Standard deviation of the regression as a function of $pK_{1,b}$ supposed. Middle: equivalence volume and $pK_{1,a}$ values as a function of $pK_{1,b}$ supposed. Bottom: plots for a series of $pK_{1,b}$ values selected. Data [*V*; *pH*]: [0.62; 7.75] [1.81; 7.70] [2.89; 7.65] [3.85;7.60] [4.71;7.55] [5.49;7.50] [6.18;7.45] [6.80;7.40] [7.36; 7.35] [7.85;7.30] [8.30; 7.25] [9.05; 7.15] [9.65; 7.05] [10.13; 6.95][10.52; 6.85][10.68; 6.80] [10.82; 6.75][11.06; 6.65][11.53; 6.35].

Figure 4 (top) shows the titration curve corresponding to the series of eleven pairs of data (left side) and the application of the conventional Gran method (right side). **Figure 4** (middle) shows the linearization method proposed in this chapter (left side) and the Gran method for data beyond the equivalence point (right side). Finally, at the bottom of **Figure 4**, a comparative study of the three methods is shown together. Gran and modified Schwartz (proposed method) on the same scale, showing the superiority of the latter transforming the whole curve in a straight line. The scale corresponding to the application of the Gran method in acid medium ($V>V_{eq}$) is shown on the secondary axis.

On the other hand, the proposed method, based on the use of Eq. (35), has been applied to a series of 19 data pairs (compiled in the legend of **Figure 5**) in the first instance, assuming the value of $pK_{1,b}$ unknown and varying its in 0.01 units of pK in the vicinity of $pK_{1,b}$ true value. The standard deviation of the corresponding regression lines ($s_{y/x}$) obtained as a function of the assumed $pK_{1,b}$ value is shown in **Figure 5** (top). The standard deviation of the regression is minimal when the value of $pK_{1,b}$ agrees on the true value (or differs in less than 0.01 unit). The values obtained for the equivalence volume and for $pK_{1,a}$ (ammonium ion), as a function of the assumed value of $pK_{1,b}$, are plotted in the central part of **Figure 5**.

Two conclusions can be drawn from the study:

1. The proposed method is sufficiently robust regarding to variations in the supposed value for $pK_{1,b}$. Values of $pK_{1,b}$ between 9.09 and 9.16 give rise to an error in the determination of the equivalence volume < 0.1%.

2. The proposed method does not work to estimate, simultaneously, the value of the acidity constant of the ammonium ion, as shown in the central part of **Figure 5** by plotting $pK_{1,a}$ obtained (secondary axis) versus the supposed value of $pK_{1,b}$. Small variations in the supposed value of $pK_{1,b}$ lead to large variations in the value obtained for $pK_{1,a}$. Moreover, in case of $pK_{1,a}$ is not applicable a minimization criterion because a minimum or maximum value is not reached when $pK_{1,b}$ agrees on the true value. When $pK_{1,b} < 9.11$, the slope of the line changes from negative to positive, so Eq. (50b) is not applicable, since the value of the constant obtained is negative, which has no physical meaning.

10. Checking the proposed linearization method with experimental data

Once tested that our proposed method works well, the theory devised has been applied to the experimental data recently reported by Cruz [68] in the Journal of Chemical Education in a study about the determination of ammonia with HCl using concentrated (4% w/v) and diluted (1.3% w/v) boric acid to reproduce the final-Kjeldahl titration when two different volumes of ammonia distillate are collected (data [*V, pH*] appear as supporting information).

Table 4 shows the data [*V, pH*] corresponding to the titration curves of ammonia in solutions of H_3BO_3 at 4% w/v (0.6469M) and 1.33% w/v (0.2156M). The shape of titration curves and the application of the conventional Gran method in the region prior to equivalence point are shown in **Figure 6**. A curvilinear shape is obtained with both solutions (being the curve most flattened when $[H_3BO_3]$=0.6469M), so that, in its original philosophy, the Gran method is not applicable.

In order to apply our proposed linearization method, data with V_{HCl} < 11mL are selected (region prior to V_{eq}, which is around 12mL). **Tables 5** and **6** show the calculations performed

V_0=50 mL; C(HCl)=0.12M						V_0=150 mL; C(HCl)=0.12M					
C(H_3BO_3)=0.6469M (4% w/v)						C(H_3BO_3)=0.2146M (1.33% w/v)					
V	pH	V	pH	V	pH	V	pH	V	pH	V	pH
0.00	6.45	11.35	4.95	11.91	4.12	0.00	7.38	11.37	5.80	12.01	4.32
2.41	6.31	11.38	4.94	11.93	4.02	1.31	7.29	11.48	5.70	12.04	4.15
3.48	6.23	11.43	4.89	11.96	3.91	2.21	7.23	11.54	5.65	12.07	4.00
4.60	6.14	11.49	4.84	11.99	3.79	3.20	7.16	11.59	5.59	12.10	3.88
5.38	6.08	11.52	4.82	12.01	3.67	4.28	7.07	11.62	5.55	12.13	3.79
6.39	5.99	11.55	4.79	12.04	3.57	5.21	7.00	11.65	5.51	12.16	3.71
7.42	5.88	11.58	4.76	12.07	3.46	6.22	6.90	11.69	5.47	12.21	3.58
8.40	5.76	11.63	4.69	12.10	3.38	7.26	6.80	11.71	5.43	12.27	3.48
9.37	5.61	11.66	4.66	12.13	3.31	8.14	6.69	11.74	5.37	12.33	3.40
10.38	5.38	11.69	4.67	12.18	3.18	9.02	6.55	11.78	5.32	12.41	3.30
10.56	5.32	11.72	4.58	12.24	3.08	9.50	6.47	11.80	5.25	12.50	3.23
10.92	5.18	11.75	4.53	12.30	3.00	10.00	6.35	11.83	5.18	12.62	3.14
11.04	5.13	11.77	4.48	12.35	2.93	10.40	6.24	11.86	5.09	12.74	3.07
11.15	5.07	11.80	4.42	12.44	2.84	10.78	6.11	11.89	4.99	12.88	2.99
11.21	5.04	11.83	4.36			10.92	6.05	11.92	4.87	13.06	2.92
11.27	5.01	11.85	4.29			11.11	5.96	11.95	4.72	13.29	2.84
11.32	4.97	11.88	4.21			11.22	5.89	11.98	4.52		

Table 4. Data [*V, pH*] corresponding to the titration curves of ammonia in solutions of H_3BO_3 at 4% w/v (0.6469M) and 1.33% w/v (0.2156M) [68].

when applying the proposed linearization method for the optimum $pK_{1,b}$ (value shown in the box) in the studies of the ammonia solutions diluted (0.2156M) and concentrated (0.6469M), respectively.

The most diluted solution ($[H_3BO_3]$=0.2156M) is firstly studied. **Figure 7** (top) shows the plots obtained for different supposed values of $pK_{1,b}$, in the range from 9.04 to 9.24. The best straight line corresponds to a value of $pK_{1,b}$ around 9.12, as shown in the bottom left side of **Figure 7** plotting the standard deviation of the regression obtained for each line versus the supposed value of $pK_{1,b}$, which reaches a minimum when $pK_{1,b}$ equals to 9.12. The supposed $pK_{1,b}$ is then varied around 9.12 in 0.001 units (**Figure 7**, bottom right), which minimizes the standard deviation of the regression to 9.117.

In the study of $[H_3BO_3]$=0.6469M, the results obtained indicate it is necessary to explore the values for supposed $pK_{1,b}$ in a wider range toward lower values of pK, consistent with the fact that the boric acid strength increases with its concentration [68, 70, 71]. It should be noted that the $pK_{1,b}$ value used with the proposed method is an adjustment parameter, and in fact

http://dx.doi.org/10.5772/intechopen.68826

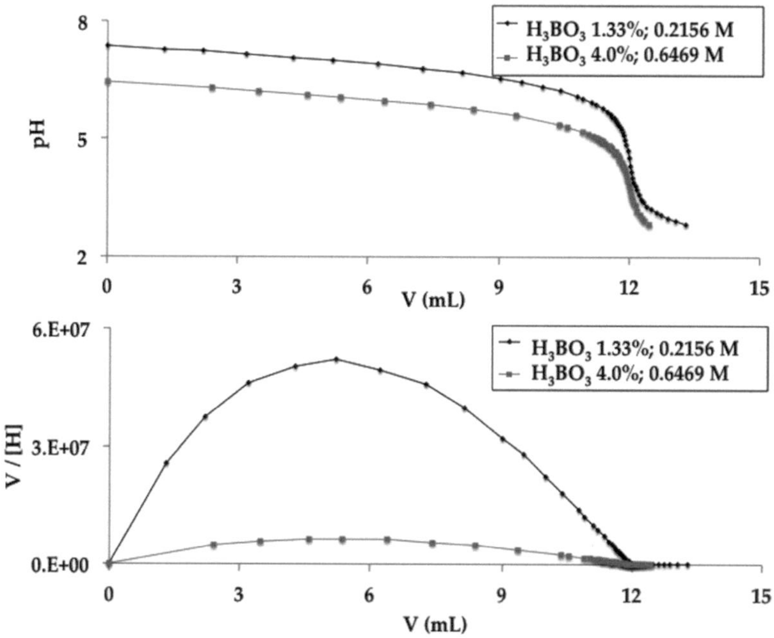

Figure 6. Ammonia titration curve ($pH=f(V)$) in boric acid (top) and conventional Gran method (bottom).

apparent, without a specific physical meaning. From **Figure 8** (top), it can be observed that when the apparent $pK_{1,b}$ reaches a value around 7.5, the representation of $V'[H]$ versus V' has the form of a straight line. This parameter is varied in 0.001 units around this value, being verified that the standard deviation of the regression is minimum (**Figure 8** bottom) when it reaches the value of 7.509 corresponding to a V_{eq} of 11.827mL.

Figure 9 shows the application of the proposed method for the ammonia solutions diluted (0.2156M) and concentrated (0.6469M) using the optimum value for $pK_{1,b}$ in each case (9.117 and 7.509, respectively). Note how the response of the diluted sample in boric acid, in a relation to 1:3, is more sensitive (greater slope) than that of the sample more concentrated in boric acid.

An empirical parameter, Γ, can be defined, which take into account the variation in the value of $pK_{1,b}$ optimum when the boric acid concentration increases, reflecting the increase in acidity motivated by the appearance of polynuclear species. The complexity of such treatment far exceeds the objective of this book chapter.

$$\Gamma = \frac{[K_{1,b}]_c}{[K_{1,b}]_d} = 10^{([K_{1,b}]_c - [K_{1,b}]_d)} = 10^{(9.117-7.509)} = 40.55 \quad (52)$$

The determination of the various polyborate species in solution has proved to be difficult because they appear at fairly high concentrations of boric acid and involve the addition of no more than one OH^- ion per $B(OH)_3$ group [87]. The pK values of triborate

V_0=	150 mL			V(eq)=	12.024 mL	
C(HCl)=	0.12M			pK(*)=	7.179	
pKa(H_3BO_3)=	9.117					
C(H_3BO_3)=	0.2156M		a_1=	-1.510E+07	1.816E+08	=a_0
pKw=	13.8		$s_{(a1)}$=	1.353E+05	1.189E+06	=$s_{(a0)}$
			R^2=	0.9990	1.015E+06	=$s_{(y/x)}$
V	**V/[H]**	**pH**	**Δ**	**$f_{0,b}$**	**V'**	**V'/(H)**
1.31	2.554E+07	7.29	-2.577E-07	1.468E-02	5.265E+00	1.027E+08
2.21	3.753E+07	7.23	-2.103E-07	1.281E-02	5.661E+00	9.614E+07
3.20	4.625E+07	7.16	-1.599E-07	1.092E-02	6.143E+00	8.880E+07
4.28	5.029E+07	7.07	-1.011E-07	8.894E-03	6.677E+00	7.845E+07
5.21	5.210E+07	7.00	-5.849E-08	7.580E-03	7.253E+00	7.253E+07
6.22	4.941E+07	6.90	0.000E+00	6.031E-03	7.845E+00	6.232E+07
7.26	4.581E+07	6.80	5.849E-08	4.796E-03	8.553E+00	5.396E+07
8.14	3.987E+07	6.69	1.265E-07	3.727E-03	9.144E+00	4.479E+07
9.02	3.200E+07	6.55	2.256E-07	2.703E-03	9.748E+00	3.459E+07
9.50	2.804E+07	6.47	2.921E-07	2.249E-03	1.011E+01	2.982E+07
10.00	2.239E+07	6.35	4.112E-07	1.707E-03	1.046E+01	2.342E+07
10.40	1.807E+07	6.24	5.479E-07	1.326E-03	1.076E+01	1.869E+07
10.78	1.389E+07	6.11	7.558E-07	9.830E-04	1.104E+01	1.423E+07
10.92	1.225E+07	6.05	8.735E-07	8.563E-04	1.115E+01	1.251E+07

Table 5. Linearization method applied to the titration data lower than 11 mL (*Veq*≈12 mL): diluted boric acid series (1.33%, 0.2156M).

($3B(OH)_3{=}B_3O_3(OH)_4^- + H^+ + 2H_2O$) and pentaborate ($5B(OH)_3{=}B_5O_6(OH)_4 + H^+ + 5H_2O$) are of the order of 6.60 and 6.90, respectively [68].

The titration with H_3BO_3 0.2156M results in a wider *pH* range and a steeper slope, contrary to that predicted by the theoretical models of the titration curves. This is for the concurrent presence of other polymer species more abundant in H_3BO_3 0.6469M, which disappear progressively as the dilution increases. The *pH* value at the equivalence point is variable depending on the initial concentration of boric acid, which can lead to systematic errors when using colored indicators in the detection of the end point.

Figure 10 (top) shows the conventional Gran method (I) to the data compiled in **Table 4**. Despite the precision of the data, the application of this method has difficulties in the appreciation of the equivalence volume. Cruz [68] applied the Gran method (II) to the titration data between 11 and 12mL, which could be questioned from the methodological point of view. The results obtained by Cruz are shown in **Figure 10** (bottom).

Table 7 summarizes a comparative study of the results obtained in the calculation of the equivalence point using the proposed method described in this book chapter and those

V_0=	50 mL			V(eq)=	11.827 mL	
C(HCl)=	0.12M					
pKa(H_3BO_3)=	7.509					
C(H_3BO_3)=	0.6469M		a_1=	5.662E+06	-6.697E+07	=a_0
pKw=	13.8		$s_{(a1)}$=	2.662E+04	3.828E+05	=$s_{(a0)}$
			R^2=	0.9998	1.749E+05	=$s_{(y/x)}$
V	**V/[H]**	**pH**	**Δ**	**$f_{0,b}$**	**V'**	**V'/(H)**
2.41	4.921E+06	6.31	4.574E-07	5.948E-02	1.844E+01	3.765E+07
3.48	5.910E+06	6.23	5.619E-07	4.997E-02	1.695E+01	2.878E+07
4.60	6.350E+06	6.14	7.026E-07	4.100E-02	1.565E+01	2.161E+07
5.38	6.468E+06	6.08	8.127E-07	3.590E-02	1.506E+01	1.810E+07
6.39	6.245E+06	5.99	1.008E-06	2.938E-02	1.431E+01	1.398E+07
7.42	5.629E+06	5.88	1.306E-06	2.296E-02	1.361E+01	1.032E+07
8.40	4.834E+06	5.76	1.729E-06	1.751E-02	1.312E+01	7.549E+06
9.37	3.817E+06	5.61	2.448E-06	1.246E-02	1.273E+01	5.185E+06
10.38	2.490E+06	5.38	4.165E-06	7.375E-03	1.237E+01	2.966E+06
10.56	2.206E+06	5.32	4.783E-06	6.430E-03	1.229E+01	2.568E+06
10.92	1.653E+06	5.18	6.605E-06	4.666E-03	1.217E+01	1.843E+06

Table 6. Linearization method applied to the titration data lower than 11 mL (*Veq*≈12 mL): concentrated boric acid series (4% w/v, 0.6469M).

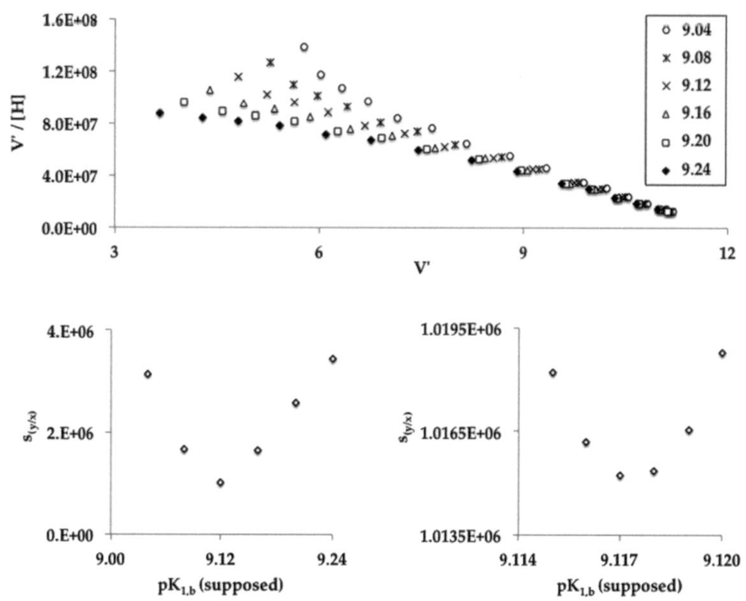

Figure 7. Top: proposed linearization method as a function of $pK_{1,b}$ supposed. Data [*V, pH*] are shown in **Table 4**. Bottom: standard deviation of the regression as a function of $pK_{1,b}$ supposed.

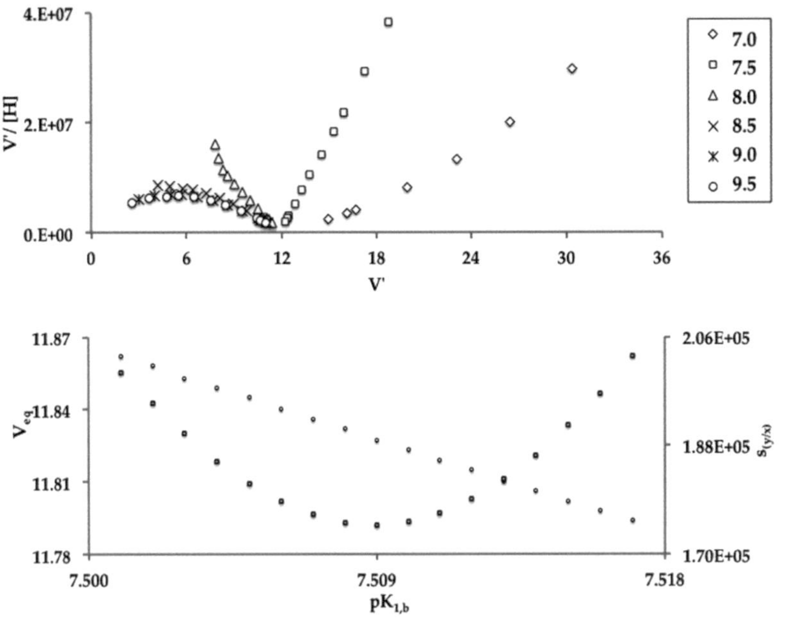

Figure 8. Top: proposed linearization method in the titration with $[H_3BO_3]$ = 0.6469 and V_0 = 50 mL when V < V_{eq}. Bottom: equivalence volume and $s_{(y/x)}$ values as a function of $pK_{1,b}$ supposed; $[H_3BO_3]$=0.6469 M.

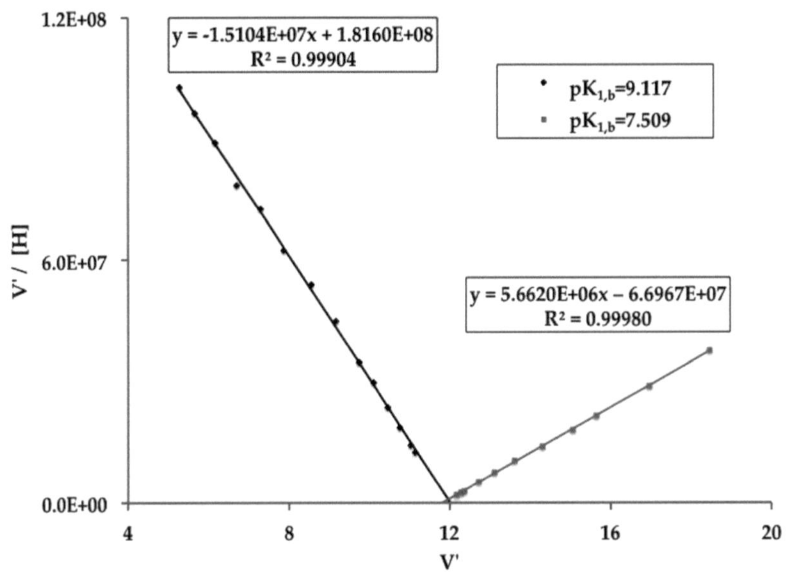

Figure 9. Left: proposed linearization method in the titration with $[H_3BO_3]$=0.2156 and V_0=150mL. Right: proposed linearization method in the titration with $[H_3BO_3]$=0.6469 and V_0=50mL.

obtained by Cruz [68]. The differences range from 0.15% for $[H_3BO_3]$=0.2156M, and −1.6% for $[H_3BO_3]$=0.6469M. The values of the standard deviation of the volume at the equivalence point are equal to 0.036 and 0.017 volume units for the proposed method for $[H_3BO_3]$ = 0.2156 M and

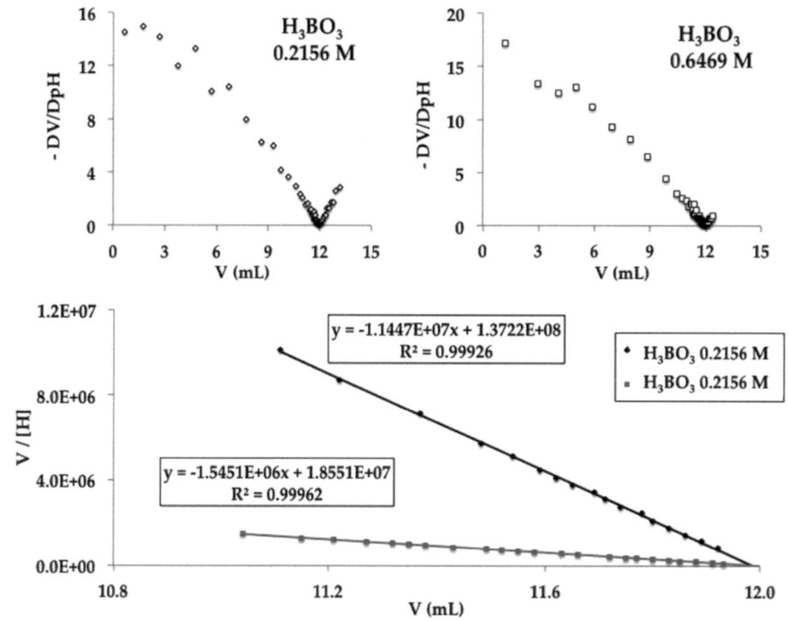

Figure 10. Top: Gran I method. Bottom: Gran II method (conventional) for $11<V<12$mL (as applied by Cruz (2013)).

$[H_3BO_3]$	pH range	N	Method	$pK_{1,b}$	V(eq)	s[V(eq)]	s[V(eq)]*
0.2156	1.31–10.92	14	Proposed	9.117	12.024	0.036	0.133
0.6164	2.41–10.92	11	Proposed	7.509	11.794	0.017	0.102
0.2156	11.04–11.93	23	Gran II	-	12.006	-	-
0.6164	11.11–11.92	17	Gran II	-	11.987	-	-

Table 7. Comparison results obtained in the calculation of the equivalence point using the proposed method and the Gran method [68].

0.6469 M, respectively, and 0.3 and 0.15% in terms of relative standard deviation, correspondingly. If the covariance is not taken into account, these values would increase in absolute terms to 0.133 and 0.102 units of volume corresponding to 1.11 and 0.86%, respectively (an increase of the relative error between four and six times more). So, the covariance of measurements can be important as the variances and both contribute significantly to the total analytical error.

Finally, the proposed method has been applied to experimental data carried out in our laboratory reproducing the conditions reported by Cruz [68] in the case of diluted boric acid.

10.1. Reagents

All reagents were analytical grade unless otherwise specified:

- Boric acid (H_3BO_3) M=61.83g/mol (Merck>99.5%), Ammonia (30%) M=17.03g/mol. (Panreac); Trishydroxymethyl aminomethane (TRIS; $C_4H_{11}NO_3$) M=121.14g/mol.

- Buffers pH=3, 5 (Riedel-de Häen), 4.01, 7 (Crison).
- Phial of HCl 1M (Riedel-de Häen).

10.2. Instruments

Analytical balance (Metler AE200) (4 decimals), Granatario (Metler PJ 400) (2 decimals), pH-meter Crison GLP 21, with a combined Ag/AgCl glass electrode were used. The pH-meter was calibrated using *pH* buffers 3, 4.01, 5 and 7, using a two-point calibration method. Burette of 5 mL (Brand) (± 0.01 at 20ºC) was used.

10.3. Experimental

About 5mL of ammonia solution (0.12M) was pipetted into a 50mL volumetric flask and made up to the mark with a boric acid solution (4% w/v). The contents of this solution were transferred to a 200mL Erlenmeyer flask containing 100mL of distilled water. Then, the ammonium borate solutions were titrated potentiometrically with HCl (0.077M) (previously standardized with TRIS) using the glass *pH* electrode.

Table 8 shows the data [*V, pH*] corresponding to the titration curve of ammonia in solution of H_3BO_3 (0.1941M). The shape of the valuation curve is shown in the upper part of **Figure 11**. All data prior to the equivalence point, before the jump in the titration curve, have been taken to apply the calculus by the proposed method, which leads to an equivalence volume equal to 6.17mL. The equivalence volume obtained by the first derivative method is equal to 6.20mL (although not many points are available in this case).

If the first points of the titration in which the solution is not sufficiently buffered, are neglected, a slightly lower volume (6.14mL) is obtained. An analogous result is obtained by using only

V	pH	V	pH	V	pH	V	pH
0.10	6.609	1.20	6.506	2.80	6.340	4.81	5.922
0.19	6.594	1.42	6.489	3.00	6.302	5.00	5.847
0.30	6.590	1.51	6.483	3.20	6.278	5.20	5.769
0.41	6.577	1.60	6.474	3.40	6.243	5.40	5.687
0.50	6.565	1.71	6.466	3.60	6.210	5.60	5.537
0.59	6.562	1.81	6.444	3.80	6.171	5.80	5.303
0.70	6.549	2.00	6.429	4.00	6.131	6.00	4.907
0.80	6.544	2.19	6.406	4.20	6.095	6.20	4.275
0.92	6.534	2.40	6.383	4.40	6.038	6.40	3.787
1.10	6.522	2.60	6.358	4.60	5.985	6.60	3.574

Table 8. Data [*V, pH*] corresponding to the titration curves of ammonia in solutions of H_3BO_3 (0.1941M) with HCl (0.077M).

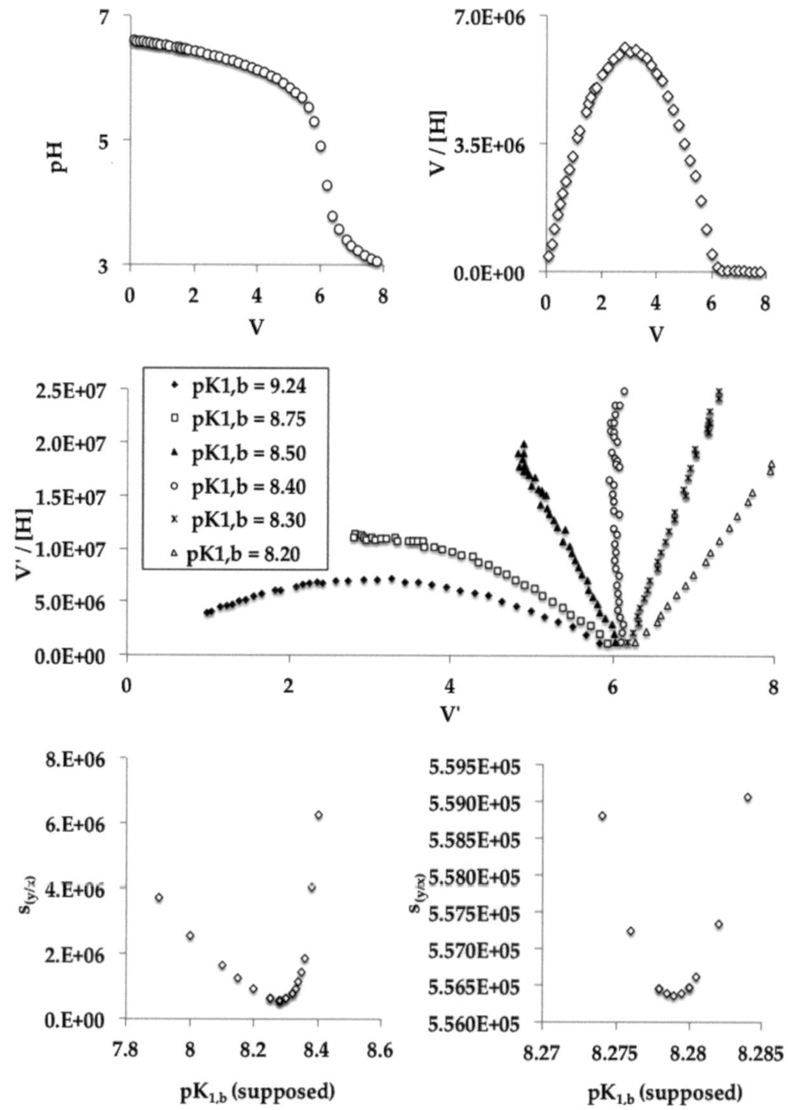

Figure 11. Top left: ammonia titration curve ($pH=f(V)$) in boric acid (0.1941 M) with HCl (0.077 M). Top right: conventional Gran method. Middle: proposed linearization method as a function of $pK_{1,b}$ supposed. Bottom: $s_{(y/x)}$ values as a function of $pK_{1,b}$ supposed.

the last 9 points prior to the equivalence point. However, the great contribution of this method is the complete linearization.

Figure 11 (middle) shows the plots obtained for different supposed values of $pK_{1,b}$, in the range from 8.20 to 9.24. The best straight line corresponds to a value of $pK_{1,b}$ around 8.279, as shown in the bottom of **Figures 11** and **12**.

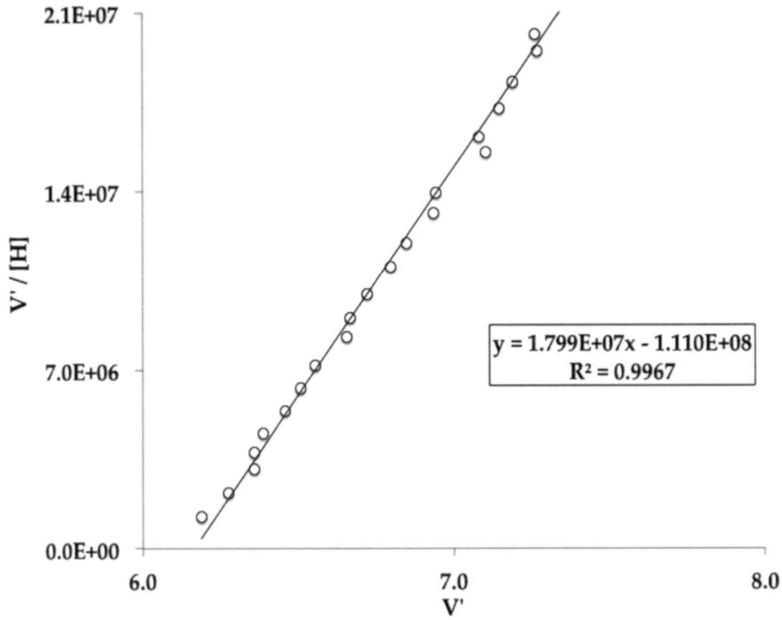

Figure 12. Proposed linearization method in the titration of ammonia in $[H_3BO_3]=0.1941$ with HCl (0.077M).

11. Conclusions

The knowledge on the mathematical approach involved in the titration curves is very useful to check the validity of the procedure. A thoroughly theoretical study of the boric acid version of the titrimetric finish of the important Kjeldahl method has been carried out in this book chapter.

In order to locate the equivalence point of the ammonia-boric acid titration, an extension of the method of Schwartz has been devised. Though the method is non-approximate, it requires the knowledge of the acidity constant of the boric acid, which may be calculated by a trial and error procedure (minimization of the standard deviation of the regression line). Unlike the differential method, the proposed method makes use of all the experimental data, so no preliminary knowledge concerning the end point is needed. The method has proved very accurate when applied to synthetic titration curves and data, and in order to check its utility, it has been applied to the experimental data recently reported by Cruz [68]. In addition, titration data have been obtained in the laboratory and processed consequently, with good results.

The study of experimental ammonia-boric acid systems titrated with hydrochloric acid allows us to extract interesting conclusions. First, the behavior of experimental systems under study is far from the expected one on the basis of the theoretical model. This difference will be greater as the concentration of boric acid used as trapping agent for ammonia increases. Second, the proposed method works well so that it allows the straight line model fits properly the experimental data and leads to a reliable equivalence end point value by using a minimization criteria as indicated above. However, the parameter varied is empirical, without any physical significance. Given the complexity of the systems with polynuclear borate species being present at high boric acid concentration, it tends to disappear by diluting the solution. It would be interesting on this respect to assay another weak acid other than boric acid as trapping agent, thus avoiding the concurrent equilibria of polyborate polymer species.

Author details

Julia Martín, Lucía Fernández Sarria and Agustín G. Asuero*

*Address all correspondence to: asuero@us.es

Faculty of Pharmacy, Department of Analytical Chemistry, The University of Seville, Seville, Spain

References

[1] Kjeldahl J. Neue methode zur Bestimmung des Stickstoffs in organischen Körpern. Z. Analytical Chemistry. 1883;**22**:366–382

[2] Kjeldahl J. New method for the determination of nitrogen. Chemistry News. 1883;**48** (1240):101–102

[3] Sella A. Classic kit: Kjeldahl flask. The Royal Society of Chemistry (http://www.rsc.org/): Chemistry World; 2008

[4] Analyst Ed. On Kjeldahl method for determination of nitrogen. Analyst. 1885;**10**(6): 127–128

[5] Burns DT. Kjeldahl, the man, the method and the Carlsberg Laboratory. Analytical Proceedings. 1984;**21**(6):210–214

[6] Blyth AW. A method of determining organic nitrogen in liquids. Analyst. 1884;**9**(7): 115–116

[7] Kebler JF. Notes on the estimation of nitrogen in nitrates by Kjeldahl's method, and an index to the literature on the estimation of nitrogen. Journal of Analaytical and Pure Chemistry. 1891;**5**:257–278

[8] Zelitch I. Hubert Bradford Vickery 1893–1978, Biographical Memoir. Washington, DC: National Academy of Sciences; 1985

[9] Möller J. Protein analysis revisited. In Focus. 2010;**34**(2):22–23

[10] Jung S, Rickert DA, Deak NA, Aldin ED, Recknor J, Johnson LA, Murphy PA. Comparison of Kjeldahl and Dumas methods for determining protein contents of soybean products. Journal of the American Oil Chemists' Society. 2003;**80**(12):1169–1173

[11] FDA. Food Labeling Health Claims: Soy Protein and Coronary Heart Diseases. Silver Spring, Md: Food and Drug Administration; Final Rule, Fed. Reg. 1999;**64**:57700–57733

[12] Sáez-Plaza P, Navas MJ, Wybraniec S, Michałowski T, García Asuero A. An overview of the Kjeldahl method of nitrogen determination. Part II. sample preparation, working scale, instrumental finish, and quality control. Critical Reviews in Analytical Chemistry. 2013;**43**:224–272

[13] Dumas J-B, Lettre de M and Dumas a M. Gay-Lussac, Sur les Procédés de l'Analyse Organique. Annales de Chimie et de Physique. 1831;**2**(47):198–215

[14] Rammelsberg C. Ueber einige Salze der Untersehwfelsäure und ihre Verbindungen mit Ammoniak. Annual Review of Physical Chemistry. (Pogg. Ann.) 1843;**58**:295–298

[15] Nessler J. Verhalten des Jodquecksilbers zu Ammoniak, und eine neue Reaktion auf Ammoniak. Chemistry Central Journal. 1856;**27**:529–541; On the behavior of iodide of mercury to ammonia, and a new reaction for ammonia. Journal für praktische Chemie. 1856;**14**(445–453):463–466

[16] Wiedemann G. Ueber ein neues Zersetzungsproduct des Hanstofts. Pharmaceutical Chemistry Journal. (Pogg. Ann.) 1849;**74**:67–84

[17] Riegler E. Eine kolorimetrische Bestimmungsmethode des Eiweisses. Analytical and Bioanalytical Chemistry. 1914;**53**:242–254

[18] Gornall AG, Bardawill CJ, David MM. Determination of serum proteins by means of biuret reagent. The Journal of Biological Chemistry. 1949;**177**(2):751–766

[19] Berthelot MPE. Violet D'aniline. Repert. Chim. Appl. 1859;**1**:284

[20] Searle PL. The berthelot or indophenol reaction and its use in the analytical chemistry of nitrogen. Analyst. 1984;**109**(5):549–568

[21] Folin O, Ciocalteu J. On tyrosine and tryptophan determinates in proteins. The Journal of Biological Chemistry. 1927;**73**(2):627–650

[22] Fraenkel-Conrat H, Cooper M. The use of dye for the determination of acid and basic groups in proteins. The Journal of Biological Chemistry. 1944;**154**(1):239–246

[23] Wu H. A new colorimeric method for the determination of plasma proteins. The Journal of Biological Chemistry. 1922;**51**(1):33–39

[24] Lowry OH, Rosebrough NJ, Farr AL, Randall RJ. Protein measurement with the folin phenol reagent. The Journal of Biological Chemistry. 1951;**193**(1):265–275

[25] Peterson GL. Review of the folin phenol protein quantitation of lowry, Rosebrough, Farr and Randall. Analytical Biochemistry. 1979;**100**(2):201–220

[26] Legler G, Muller-Platz CM, Mentges-Hettkamp M, Pflieger G, Julicj E. On the chemical basis of the lowry protein determination. Analytical Biochemistry. 1985;**150**(2):278–287

[27] Owusu-Apenten RK. Kjeldahl method, Quantitative Aminoacid Analysis and Combustion Analysis. In Food Protein Analysis. Quantitative Effects on Processing; New York-Basel: Marcel Dekker; 2002. Chapter1, pp. 1–45

[28] Chandrarajan J. Klein L. Lowry assay of dilute protein solutions containing high concentrations of Triton X-100. Analytical Biochemistry. 1975;**69**(2):632–636

[29] Bradford MM. A rapid and sensitive method for the quantitation of microgram quantities of protein utilizing the principle of protein–dye binding. Analytical Biochemistry. 1976;**72** (1–2):248–254

[30] Compton SJ, Jones CG. Mechanism of dye response and interference in the bradford protein assay. Analytical Biochemistry. 1985;**151**(2):369–374

[31] Smith PK, Krohn RI, Hermanson GT, Malla AK, Gartner FH, Provenzano MD, Fujimoto EK, Goecke NM, Olson BJ, Klenk DC. Measurement of protein using bicinchoninic acid. Analytical Biochemistry. 1985;**150**(1):76–85; Erratum, 1987;**163**(1):279

[32] You WW, Haugland RP, Ran DK, Haugland RP. 3-(4-Carboxy Benzyl) Quinoline-2-Carboxaldehyde, a reagent with broad dynamic range for the assay of proteins and lipoproteins in solution. Analytical Biochemistry. 1997;**244**(2):277–282

[33] Möller J. Kjeldahl-still going strong. In Focus. 2009;**33**(1):14–16

[34] Official Methods of Analysis of AOAC International 18th Ed. (Revision 4). Gaithersburg, MD: AOAC; 2011

[35] McKenzie HA. The Kjeldahl determination of nitrogen: Retrospect and prospect. Trends in Analytical Chemistry. 1994;**13**(4):138–144

[36] Stephen WI. Determination of nitrogen in organic compounds in the years before Kjeldahl's method. Analytical Proceedings. 1984;**21**:215–220

[37] Johannsen W. Johan Kjeldahl 1849–1900. Obituary. Ber. Dtsch. Chem. Ges. 1900;**33**(3): 3881–3888

[38] Johannsen W. Johan Kjeldahl 1849–1900. CR Trav. Lab. Carlsb. (Edition Française). 1903;**5**: I-VIII

[39] Holter H, Møller KM. The Carlsberg Laboratory 1876/1976. Rhodos, Denmark: The Carlsberg Foundation; 1976

[40] Veibel S. John Kjeldahl (1849–1900). Journal of Chemical Education. 1949;**26**(9):459–461

[41] Hansen EH. The current status of analytical chemistry in Denmark. A personnel view. Analytical Proceedings. 1993;**30**(1):3–9

[42] Rideal EK. Soren Peter Lauritz Sörensen. Journal of the Chemical Society. 1940;554–561

[43] Asuero AG, Michalowski T. Comprehensive formulation of titration curves for complex acid-base systems and its analytical implications. Critical Reviews in Analytical Chemistry. 2011;**41**:151–187

[44] Michalowski T, Asuero AG. New approaches in modeling the carbonate alkalinity. Critical Reviews in Analytical Chemistry. 2012;**42**(3):220–244

[45] Breuer PL, Sutchiffe CA, Meakin RL. Cyanide measurement by silver nitrate titration: Comparison of rhodanine and potentiometric end points. Hydrometallurgy. 2011;**106**: 135–140

[46] Felber H, Rezzonico S, Máriásy M. Titrimetry at a metrological level. Metrologia. 2003;**40**: 249–254

[47] Thordarson P. Determining association constants from titration experiments in supramolecular chemistry. Chemical Society Reviews. 2011;**40**:1305–1323

[48] Michalowski T, Asuero AG, Ponikrar-Svet M, Toporek M, Pietrzyk A, Rymanowski M. Liebig-Denigès method of cyanide determination: A comparative study of two approaches. Journal of Solution Chemistry. 2012;**41**:1224–1239

[49] Michalowski T, Toporek M, Michalowska-Kaczmarcyk AM, Asuero AG. New trends in studies on electrolytic systems. Electrochimica Acta. 2013;**109**:519–531

[50] Winkler-Oswatistsch R, Eigen M. The art of titration. From classical end points to modern differential and dynamic analysis. Angewandte Chemie International Edition. 1979;**18**:20–49

[51] Sneddon J, Hardaway C, Bobbadi KK, Reddy AK. Sample preparation of solid samples for metal determination by atomic spectroscopy–An overview and selected recent applications. Applied Spectroscopy Reviews. 2006;**41**(4):1–14

[52] Brayton SV. Acid Digestions using the Hach Digesdahl Digestion Apparatus: Sample Preparation for Protein and Elemental Analysis. Hach Company, Loveland, CO: Tech. Info. Ser. Booklet; No 14; 1992

[53] Flores EMM, Barin JS, Mesko MF, Knapp G. Sample preparation tehniques based on combustion reactions in closed vessels–A brief overview and recent applications. Spectrochimica Acta Part B. 2007;**62**(9):1051–1064

[54] Egli H. Kjeldahl Guide, BUCHI. CH-9230 Flawill, Switzerland: Büchi Labortechnik AG; 2008

[55] Thiex NJ, Manson H, Anderson S, Persson JA. Determination of crude protein in animal feed, forage, grain, and oilseeds by using block digestion with a copper catalyst and steam distillation into boric acid: Collaborative study. Journal of AOAC INTERNATIONAL Int. 2002;**85**(2):309–317

[56] EN ISO 5983–1. Animal feeding stuffs–Determination of nitrogen content and calculation of crude protein content–Part 1: Kjeldahl method. Genève: International Organization for Standardization; 2009

[57] Lynch JM, Barbano DM. Kjeldahl nitrogen analysis as a reference method for protein determination in dairy products. Journal of AOAC INTERNATIONAL. 1999;**82**(6):1389–1398

[58] AMC. Methods for the destruction of organic matter. Analyst. 1960;**85**(9):643–656

[59] Greenfield H, Southgate DAT. Food composition data. Production, management and use. 2nd ed. Rome: FAO; Review of Methods of Analysis; 2003. Chapter 7; pp. 97–148

[60] Persson JA, Wennerholm M, O'Halloran S. Handbook for Kjeldahl digestión. Hilleroed: Denmark FOSS: DK-3400; 2008

[61] Thiex NJ, Manson H. New AOAC standard for crude protein determinations in animal feed, grain, and oilseed using Tecator Kjeltec equipment, In Focus. 2002;**26**(2):10–12

[62] Saha UK, Sonon L, Kissel DE. Comparison of conductimetric and colorimetric methods with distillation titration method of analyzing ammonium-nitrogen in total Kjeldahl digests. Communications in Soil Science and Plant Analysis. 2012;**43**(18):2323–2341

[63] Winkler JW. Beitrag zur titrimetrischen bestimmung des ammoniaks. Angewandte Chemie International Edition. 1913;**26**(31):231–232

[64] Beljkaš B, Matić AJ, Milovanović I, Jovanov P, Mišan A, Šaric L. Rapid method for determination of protein content in cereals and oilseeds: Validation, measurement uncertainty and comparison with the Kjeldahl method. Accreditation and Quality Assurance. 2010;**15**(10):555–561

[65] Blaedel WJ, Meloche VW. Elementary quantitative analysis: Theory and practice. Evanston, IL: Row, Petterson and Co; 1957

[66] Bradstreet RB. The Kjeldahl Method for Organic Nitrogen. London: Academic Press; 1965

[67] Laitinen HA, Harris WE. Chemical Analysis. 2nd ed. New York: McGraw-Hill; 1975. p. 115

[68] Cruz G. Boric acid in Kjeldahl analysis. Journal of Chemical Education. 2013;**90**:1645–1648

[69] Ingri N. Equilibrium studies of polyanions. On the first equilibrium steps in the acidification of $B(OH)_4^-$. An application of the self medium method. Acta Chemica Scandinavica. 1963;**17**:573–580

[70] Prideaux BR. Note on the neutralization curve of boric acid. Transactions of the Faraday Society. 1915;**11**:76–78

[71] Thygesen JE. Über die Selbstkomplexbildung der Borsäure. Zeitschrift für anorganische und allgemeine Chemie. 1938;**237**(1):101–112

[72] Kantowitz BH, Roediger H, Elmes DG. Experimental Psicology. 9thed. Belment, CA: Wadsworth; 2009

[73] Gran G. Determination of the equivalent point in potentiometric titrations. Acta Chemica Scandinavica. 1950;**4**:559–577

[74] Gran G. Determination of the equivalence point in potentiometric titrations. Part II. Analyst. 1952;**77**:661–671

[75] Gran G. Equivalence volumes in potentiometric titrations. Analytica Chimica Acta 1988; **206**:111–123

[76] Schwartz LM. Advances in acid-base Gran plot methodology. Journal of Chemical Education. 1987;**64**(11):947–950

[77] de Levie R. Principles of Quantitative Chemical Analysis. New York: McGraw-Hill; 1997

[78] de Levie R. Aqueous Acid-Base Equilibria and Titrations. Oxford: Oxford University Press, 1999

[79] Asuero AG, Gonzalez G. Fitting straight lines with replicated observations by linear regression. III weighting data. Critical Reviews in Analytical Chemistry. 2007;**37**(3): 143–172

[80] Asuero AG, Bueno JM. Fitting straight lines with replicated observations by linear regression. IV. Transforming data. Critical Reviews in Analytical Chemistry. 2011;**41**(1):36–69

[81] Ivaska A and Wanninen E. Potentiometric titration of weak acids. Analytical Letters. 1973;**6**:961–967

[82] Michalowski T, Asuero AG, Wybraniec S. The titration in the kjeldahl method of nitrogen determination: Base or acid as titrant. Journal of Chemical Education. 2013;**90**(2):191–197

[83] Asuero AG. Buffer capacity of a polyprotic acid: First derivative of the buffer capacity and pKa values of single and overlapping equilibria. Critical Reviews in Analytical Chemistry. 2007;**37**(4):269–301

[84] Asuero AG. Inflexion points of the titration curve for two step overlapping equilibria. Microchemical Journal. 1993;**47**:386–398

[85] Harris DC. Quantitative Chemical Analysis. 8th ed. New York: W.H. Freeman and Company; 2010

[86] Schwartz LM. Uncertainty of a Titration equivalence point. A graphical method using spreadsheets to predict values and detect systematic errors. Journal of Chemical Education. 1992;**69**(11):879–883

[87] Baes CF, Mesmer RE. The Hydrolysis of Cations. Malabar, FL: Krieger; 1976. pp. 104–111

Chapter 3

Intersecting Straight Lines: Titrimetric Applications

Julia Martin, Gabriel Delgado Martin and
Agustin G. Asuero

Additional information is available at the end of the chapter

http://dx.doi.org/10.5772/intechopen.68827

Abstract

Plotting two straight line graphs from the experimental data and determining the point of their intersection solve a number of problems in analytical chemistry (i.e., potentiometric and conductometric titrations, the composition of metal-chelate complexes and binding interactions as ligand-protein). The relation between conductometric titration and the volume of titrant added leads to segmented linear titration curves, the endpoint being defined by the intersection of the two straight line segments. The estimation of the statistical uncertainty of the end point of intersecting straight lines is a topic scarcely treated in detail in a textbook or specialized analytical monographs. For this reason, a detailed treatment with that purpose in mind is addressed in this book chapter. The theoretical basis of a variety of methods such as first-order propagation of variance (random error propagation law), Fieller's theorem and two approaches based on intersecting confidence bands are explained in detail. Several experimental systems described in the literature are the subject of study, with the aim of gaining knowledge and experience in the application of the possible methods of uncertainty estimation. Finally, the developed theory has been applied to the conductivity measurements in triplicate in the titration of a mixture of hydrochloric acid and acetic acid with potassium hydroxide.

Keywords: titrimetric, straight lines, breakpoint

1. Introduction

Titrimetry is one of the oldest analytical methods [1], and it is still found [2–4] in a developing way. It plays an important role in various fields as well as routine studies [5–9], being used widely in the analytical laboratory given their simplicity, speed, accuracy, good reproducibility, and low cost. It is, together with gravimetry, one of the most used methods to determine chemical composition on the basis of chemical reactions (primary method).

Independent values of chemical quantities expressed in SI units are obtained through gravimetry and titrimetry (classical analysis).

In titrimetry, the quantity of tested components of a sample is assessed by the use of a solution of known concentration added to the sample, which reacts in a definite proportion. To identify the stoichiometric point, where equal amounts of titrant react with equal amounts of analyte, indicators are used in many cases to point out the end of the chemical reaction by a color change.

Information on reaction parameters is usually obtained from an analysis of the shape of the titration curve, whose shapes depend on some factors such as the reaction of titration, the monitored specie (indicator, titrant, analyte or formed product) as well as the chosen [10, 11] instrumental technique (**Table 1**) i.e., spectrophotometry, conductimetry or potentiometry, for instance. The importance of titrimetric analysis has increased with the advance of the instrumental method of end point detection.

Linear response functions are generally preferred, and when the response function is nonlinear, a linearization procedure has been commonly used with a suitable change of variables. Plotting two straight lines graphs from the experimental data and determining the point of their intersection solve a number of problems in analytical chemistry [10, 11] (**Table 1**). In segmented linear titration curves, the end point is defined by the intersection of the two straight segments. In some common examples in analytical chemistry (conductometric, spectrophotometric and amperometric titrations), this intersection lies beyond the linear ranges, and deviations from linearity are often observed directly at the end point. All curvature points should be excluded from the computation. The accuracy and precision of the results of a titrimetric determination are influenced not only by the nature of the titration reaction but also by the technique [10, 11] of the end-point location.

The problem of finding the breakpoint of two straight lines joined at some unknown point has a long statistical history [12–14] and has received considerable attention in the statistical literature. The problem in question is known by a variety of names (**Table 2**) [12–27]. Computer analysis [28–31], elimination of outliers [32, 33], and confidence limits for the abscissa [22, 34, 35] have been subject to study.

At the point of intersection (x_I), the two lines have the same ordinate. The estimation of statistical uncertainty of end points obtained from linear segmented titration is the subject

Technique	Measured property
Conductimetric titrations	Electrical conductivity
Potentiometric titrations	Potential of an indicator electrode
Spectrophotometric titrations	Absorbance
Amperometric titrations	Diffusion current at a polarizable indicator (dropping mercury or rotating platinum) electrode

Table 1. Instrumental end point detection techniques more widely applied.

Name	Authors
Breakpoint	Jones and Molitoris [12]; Shanubhogue et al. [15]
Changepoint	Csörgo and Horváth [16]; Krishanaiah and Miao [17]
Common intersection point	Rukhin [18]
Hockey stick regression	Yanagimoto and Yamamoto [19]
Intersystem crossing	Kita et al. [20]
Piecewise linear regression	Vieth [21]
Segmented regression	Piegorsch [22]
Transition	Bacon and Watts [23]
Two phase linear regression	Christensen [24]; Lee et al. [25],; Seber [26, 27]; Shaban [14]; Sprent [13]

Table 2. Names received in the literature for the intersecting point of two straight lines.

of this chapter. The topic is scarcely treated in [36, 37] analytical monographs. The method of least squares is the most common and appropriate choice and when the relative statistical uncertainties of the x data are negligible compared to the y data. Single linear regression or weighted linear regression may be applied depending on whether the variance of y is constant or varies from point to point with the magnitude of the response y, respectively.

The theoretical basis of a variety of method such as first-order propagation of variance for the abscissa or intersection, the application of Fieller's method [38–43], and other methods based on intersecting hyperbolic confidence bands as weighted averages [57, 58] of the abscissas of the confidence hyperbolas at the ordinate of intersection will be dealt in detail in this book chapter. In addition, several experimental systems will be the subject of study, with the aims of gaining knowledge and experience in the application of these methods to uncertainty estimation.

2. Theory

V-shaped linear titration curves (**Table 1**) are well known in current analytical techniques such as conductimetry, radiometry, refractometry, spectrophotometric and amperometric titrations as well as in Gran's plot. In this kind of titrations, the end point is usually located at the intersection of two lines when a certain property (conductance, absorbance, diffusion current) is plotted against the volume x of titrant added to the unknown sample containing the analyte.

Let N_1 observations on the first line

$$\hat{y}_1 = a_1 + b_1 x \tag{1}$$

and N_2 observations on the second

$$\hat{y}_2 = a_2 + b_2 x \tag{2}$$

where a_1, b_1, a_2, b_2 are the usual least squares estimates of the kth line ($k = 1, 2$), respectively. As it is stated in the introduction section when the relative statistical uncertainties of the x data are negligible compared to the y data, the use of the least squares method is the most common alternative. The ordinate variance can be considered on a priori grounds to vary systematically as a function of the position along the curve, so that weighted least squares analysis is appropriate. Formulae for calculating the intercept a, the slope b and their standard errors by weighted linear regression [59] are given in **Table 3**, where the analogy with simple linear regression (i.e., $w_i = 1$), is evident.

Note that in summation (1) and (2) by dividing by N_1 and N_2, respectively, we get

$$\overline{y}_1 = a_1 + b_1 \overline{x}_1 \tag{3}$$

$$\overline{y}_2 = a_2 + b_2 \overline{x}_2 \tag{4}$$

At the point of intersection, the lines (1) and (2) have the same ordinate $\hat{y}_1 = \hat{y}_2$ and the abscissa of intersection (denotes by $\hat{x}_I$) is given by

$$a_1 + b_1 \hat{x}_I = a_2 + b_2 \hat{x}_I \tag{5}$$

$$\hat{x}_I = \frac{a_2 - a_1}{b_1 - b_2} = -\frac{\Delta a}{\Delta b} \tag{6}$$

Random error in the points produces uncertainty in the slopes and intercepts of the lines, and therefore in the point of intersection. The probability that a confidence interval contains the true value is equal to the confidence level (e.g., 95%).

- Equation
  $\hat{y}_i = a + bx_i$
- Weights
  $w_i = 1/s_i^2$
- Explained sum of squares
  $SS_{Reg} = \sum w_i(\hat{y}_i - \overline{y})^2$
- Residual sum of squares
  $SSE = \sum w_i(y_i - \hat{y}_i)^2$
- Mean
  $\overline{x} = \sum w_i x_i / \sum w_i$
  $\overline{y} = \sum w_i y_i / \sum w_i$
- Sum of squares about the mean
  $S_{XX} = \sum w_i(x_i = \overline{x})^2$
  $S_{YY} = \sum w_i(y_i = \overline{y})^2$
  $S_{XY} = \sum w_i(x_i = \overline{x})(y_i - \overline{y})$
- Slope
  $b = S_{XY}/S_{xx}$
- Intercept
  $a = \overline{y} - b\overline{x}$
- Weighted residuals
  $w_i^{1/2}(y_i - \hat{y}_i)$
- Correlation coefficient
  $r = S_{XY}/\sqrt{S_{XX}S_{YY}}$
- Standard errors
  $s_{y/x}^2 = \frac{SSE}{n-2} = \frac{S_{YY} - b^2 S_{XX}}{n-2}$
- $s_a^2 = s_{y/x}^2 \left(\sum w_i x_i^2\right) / \left(S_{XX} \sum w_i\right)$
- $s_b^2 = s_{y/x}^2 / S_{XX}$
- $Cov(a, b) = \overline{x} s_{y/x}^2 / S_{XX}$

Table 3. Formulae for calculating statistics for weighted linear regression.

3. First-order propagation of variance for $V[\hat{x}_I]$

The precision of the point of intersection and the corresponding statistical confidence interval can be found in the simplest way by considering the random error propagation law [60]. Some authors [61, 62] evaluate the uncertainty in $\hat{x}_I$ on this way. First-order propagation of variance retains only first derivatives in the Taylor expansions and this procedure leads to

$$\begin{aligned} V[\hat{x}_I] &= \left(\frac{\partial \hat{x}_I}{\partial \Delta a}\right)^2 V[\Delta a] + \left(\frac{\partial \hat{x}_I}{\partial \Delta b}\right)^2 V[\Delta b] + 2\left(\frac{\partial \hat{x}_I}{\partial \Delta a}\right)\left(\frac{\partial \hat{x}_I}{\partial \Delta b}\right) Cov(\Delta a, \Delta b) \\ &= \left(\frac{\partial\left(\frac{-\Delta a}{\Delta b}\right)}{\partial \Delta a}\right)^2 V[\Delta a] + \left(\frac{\partial\left(\frac{-\Delta a}{\Delta b}\right)}{\partial \Delta b}\right)^2 V[\Delta b] + 2\left(\frac{\partial\left(\frac{-\Delta a}{\Delta b}\right)}{\partial \Delta a}\right)\left(\frac{\partial\left(\frac{-\Delta a}{\Delta b}\right)}{\partial \Delta b}\right) Cov(\Delta a, \Delta b) \\ &= \frac{V[\Delta a]}{\Delta b^2} + \frac{\Delta a^2 V[\Delta b]}{\Delta b^4} - 2\frac{\Delta a Cov(\Delta a, \Delta b)}{\Delta b^3} \end{aligned} \tag{7}$$

valid in those cases, in which the standard deviations of the ordinate data are a small fraction of their magnitude. Taking into account Eq. (6), Eq. (7) may be rewritten as follows

$$\begin{aligned} V[\hat{x}_I] &= \frac{1}{\Delta b^2}\left(V[\Delta a] + \frac{\Delta a^2}{\Delta b^2} V[\Delta b] - 2\frac{\Delta a}{\Delta b} Cov(\Delta a, \Delta b)\right) \\ &= \frac{V[\Delta a] + \hat{x}^2{}_I V[\Delta b] + 2\hat{x}_I Cov(\Delta a, \Delta b)}{\Delta b^2} \end{aligned} \tag{8}$$

Then, the standard error estimate of $\hat{x}_I$ is as follows

$$s(\hat{x}_I) = \sqrt{V[\hat{x}]} \tag{9}$$

The end point $\hat{x}_I$ depends on four least squares parameters a_1, a_2, b_1, b_2 that are random variables. Segment one parameters depend only on measurements made along segment one and these are statistically independent of the measurements along segment two. However, Δa, and Δb are correlated random variables because each involves b_1 and b_2. Note that, a_1 and a_2 are related to b_1 and b_2 by means of Eqs. (1) and (2).

The variances of Δa and Δb are given by

$$V[\Delta a] = V[a_1 - a_2] = V[a_1] + V[a_2] \tag{10}$$

$$V[\Delta b] = V[b_1 - b_2] = V[b_1] + V[b_2] \tag{11}$$

and for the covariance between Δa and Δb, we get [63]

$$\begin{aligned} Cov(\Delta a, \Delta b) &= Cov(a_1 - a_2, b_1 - b_2) \\ &= Cov\left(\overline{y}_1 - b_1\overline{x}_1 - (\overline{y}_2 - b_2\overline{x}_2), b_1 - b_2\right) = Cov(\overline{y}_1 - \overline{y}_2 - b_1\overline{x}_1 + b_2\overline{x}_2, b_1 - b_2) \\ &= Cov(\overline{y}_1 - \overline{y}_2, b_1 - b_2) - Cov(b_1\overline{x}_1 - b_2\overline{x}_2, b_1 - b_2) \\ &= -Cov(b_1\overline{x}_1 - b_2\overline{x}_2, b_1 - b_2) \\ &= -\overline{x}_1 V[b_1] - \overline{x}_2 V[b_2] \end{aligned} \tag{12}$$

It should be noted that in the calculations, the variance regression estimates from both line segments are pooled into a single $s_p^{\,2}$, by using the following formula which weights each contribution according to the corresponding [64–67] degrees of freedom

$$s_p^2 = \frac{\sum_{i=1}^{N_1}(y_{1i}-\bar{y}_1)^2 + \sum_{i=1}^{N_2}(y_{2i}-\bar{y}_2)^2}{(N_1-2)+(N_2-2)} = \frac{(N_1-2)s_1^2+(N_2-2)s_2^2}{N_1+N_2-4} \qquad (13)$$

The standard deviation in Eq. (13) is calculated on the assumption that the $s_{(y/x)}$ values for the two lines are sufficiently similar to be pooled.

From expression in **Table 3** for the variance of the intercept ($s_a^{\,2} = V[a]$), we may derive

$$V[a] = s^2\left[\frac{\sum w_i x_i^2}{S_{XX}\left(\sum w_i\right)}\right] = s^2\left[\frac{S_{XX}+\frac{\left(\sum w_i x_i\right)^2}{\sum w_i}}{S_{XX}\left(\sum w_i\right)}\right] = s^2\left[\frac{S_{XX}+\left(\sum w_i\right)\left(\frac{\sum w_i x_i}{\sum w_i}\right)^2}{S_{XX}\left(\sum w_i\right)}\right] \qquad (14)$$

$$= s^2\left[\frac{S_{XX}+\left(\sum w_i\right)\bar{x}^2}{S_{XX}\left(\sum w_i\right)}\right] = s^2\left[\frac{1}{\sum w_i}+\frac{\bar{x}^2}{S_{XX}}\right]$$

in which s^2 is $s_{(y/x)}$ in **Table 3**; $\sum w_i$ is the sum of weights, which simply reduces to N, the number of points if the non-weighted least squares analysis is used. Taking into account Eq. (14), Eqs. (10) and (11) lead to

$$V[\Delta a] = V[a_1] + V[a_2] = \left[\frac{1}{\left(\sum w\right)_1}+\frac{\bar{x}_1^2}{(S_{XX})_1}\right]s_1^2 + \left[\frac{1}{\left(\sum w\right)_2}+\frac{\bar{x}_1^2}{(S_{XX})_2}\right]s_2^2 \qquad (15)$$

$$V[\Delta a] = \left[\frac{1}{\left(\sum w\right)_1}+\frac{1}{\left(\sum w\right)_2}+\frac{\bar{x}_1^2}{(S_{XX})_1}+\frac{\bar{x}_1^2}{(S_{XX})_2}\right]s_p^2 \qquad (16)$$

$$V[\Delta b] = V[b_1] + V[b_2] = \frac{s_1^2}{(S_{XX})_1}+\frac{s_1^2}{(S_{XX})_2} = \left[\frac{1}{(S_{XX})_1}+\frac{1}{(S_{XX})_2}\right]s_p^2 \qquad (17)$$

$$Cov(\Delta a, \Delta b) = -\bar{x}_1 V[b_1] - \bar{x}_2 V[b_2] = -\left[\frac{\bar{x}_1}{(S_{XX})_1}+\frac{\bar{x}_2}{(S_{XX})_2}\right]s_p^2 \qquad (18)$$

Once the values of $V[\Delta a]$, $V[\Delta b]$ and $Cov[\Delta a,\Delta b]$ are known from Eqs. (16), (17) and (18), respectively, the estimate of the variance of the intersection abscissa of the two straight lines, $V[\hat{x}_I]$, is calculated by applying Eq. (8).

http://dx.doi.org/10.5772/intechopen.68827

4. Confidence interval on the abscissa of the point of intersection of two fitted linear regressions

The use of confidence intervals is another alternative to express the statistical uncertainty of $\hat{x}_I$. This method depends on the distribution function of the random variable $\hat{x}_I$. If the ordinates y_i are assumed to have Gaussian (normal) distribution, the least squares parameters as well as Δa, and Δb are also normally distributed [68]. However, $\hat{x}_I$, which even is regarded as the ratio of two normally distributed variables, is not normally distributed and, indeed, becomes more and more skewed [69] as the variance levels increase. For sufficiently small variance though, $\hat{x}_I$, is approximately normally distributed. Under these circumstances, confidence intervals may be calculated from the standard deviation of $\hat{x}_I$, which is also accurate only when variances are small.

However, the construction of the confidence interval (limits) for the equivalence point by using the Student's *t*-test

$$\hat{x}_I \pm t_{\alpha/2} s(\hat{x}_I) \tag{19}$$

where $t_{\alpha/2}$ is the Student's *t* statistics at the $1 - \alpha$ confidence level (i.e., leaving an area of $\alpha/2$ to the right) and for the number of degrees of freedom (N_1+N_2 − 4) inherent in the standard deviation of $\hat{x}_I$, could be misleading. Note that because $\hat{x}_I$ involves the ratio of random variables, first-order propagation of variance is not exact [69]. Evidently, $\hat{x}_I$ is a random variable not normally distributed unless $s(\hat{x}_I)$ is small enough. When the variances of the responses are not necessarily small, a solution to this problem is to apply the called Fieller's theorem [38–43]. Another point of view is focused on the problem in the calculation of the limits of the confidence intervals by using the confidence bands for the two segmented branches.

5. The Fieller's theorem

This theorem [38–43] is supported by two capital premises:

i. Any linear combination z of normally distributed random variables is itself normally distributed.

ii. If the standardized variable $\frac{z}{\sqrt{V[z]}}$ is distributed as $N(0, 1)$, then z is distributed as t.

Consider now any pair of individual line segments written as a difference z as follows

$$z = [a_1 + b_1 x_I] - [a_2 + b_2 x_I] = \Delta a + x_I \Delta b \tag{20}$$

Note that for any such pair of lines, the difference z is not, in general, zero, because the "best" end point cannot be the one for each pair of lines of the collection. However, the mean $\langle z \rangle$ of all these z values is zero and z are normally distributed because it is formed as a linear combination of normally distributed variables. Taking into account that a_1, a_2, b_1 and b_2 are normally distributed, then z will be normally distributed. Then, in the vicinity of the intersection point, z has zero mean and its variance is

$$V[z] = V[\Delta a + x_I \Delta b] = V[\Delta a] + x_I^2 V[\Delta b] + 2x_I Cov(\Delta a, \Delta b) \tag{21}$$

and therefore, $\frac{z}{\sqrt{\hat{V}[z]}}$ is distributed as $N(0, 1)$ and according to (ii)

$$\frac{z}{\sqrt{\hat{V}[z]}} = t \tag{22}$$

This is called Fieller's Theorem [34, 38]. The development of Eq. (11) leads to the equation

$$\frac{(\Delta a + x_I \Delta b)^2}{V[\Delta a] + x_I^2 V[\Delta b] + 2x_I Cov(\Delta a, \Delta b)} = t^2 \tag{23}$$

which on rearrangement leads to

$$(\Delta a)^2 + 2x_I \Delta a \Delta b + x_I^2 (\Delta b)^2 = t^2 V[\Delta a] + t^2 x_I^2 V[\Delta b] + 2t^2 x_I Cov(\Delta a, \Delta b) \tag{24}$$

which may be factored as

$$\left((\Delta a)^2 - t^2 V[\Delta a]\right) + 2x_I \left(\Delta a \Delta b - t^2 Cov(\Delta a, \Delta b)\right) + x_I^2 \left((\Delta b)^2 - t^2 V[\Delta b]\right) = 0 \tag{25}$$

The solution of Eq. (25) gives the confidence limits for x_I estimated, where $t_{\alpha/2}$ is the appropriate value of the Student distribution at a α significance level (confidence level $1 - \alpha$) for $N_1 + N_2 - 4$ degrees of freedom. Note that in Eqs. (21), (23), (24) and (25), the corresponding values of $V[\Delta a]$, $V[\Delta b]$ and $Cov[\Delta a, \Delta b]$ are given by Eqs. (16), (17) and (18), respectively, as in the first-order propagation of variance for $V[\hat{x}_I]$.

The first and last groups of symbols enclosed in braces in Eq. (25) has the form of hypothesis tests, that is, two-tailed tests, for significant difference of intercepts and significant difference of slopes, respectively. When the hypothesis test for different slope fails, the coefficient of x_I^2

Topic	Reference
Arrhenius plot	Cook and Charnock [44]; Han [45]; Puterman et al. [46]
Calibration curves	Baxter [47]; Bonate [48]; Mandel y Linning [49]; Schwartz [50–52]
Estimation of safe doses	Yanagimoto and Yanamoto [19]
Estimation of uncertainty in binding constants	Almansa López et al. [53]
Models for biologic half-life data	Lee et al. [25]
Position and confidence limits of an extremum	Asuero and Recamales [54]; Heilbronner [55]
Standard addition method	Franke et al. [56]

Table 4. Some applications of Fieller theorem in analytical chemistry.

becomes negative finding two complex roots [22], so Fieller confidence interval embraces the entire x-axis (the lower and upper limits should strictly be set to $-\infty$ and ∞, respectively) at the chosen level of confidence.

This method has been extensively described in some other contexts in analytical and chemical literature (**Table 4**).

6. Use of hyperbolic confidence bands for the two linear branches

Several procedures dealing with hyperbolic confidence bands approximate them by straight lines and give symmetric confidence intervals for estimated x_I [58, 61, 70–72]. Evidently, the best confidence interval would be obtained by the projection on the abscissa of the surface between the four hyperbolic arcs [73].

Because a confidence band, bounded by two hyperbolic arcs, is associated with each regression line, it is obvious that the point of intersection, x_I, is only a mean value, with which a certain confidence interval is associated. If the signal values both before and after the point of intersection are normally distributed around the line with a constant standard deviation, the point of intersection and its statistical confidence interval will be estimated by the projection of the intersection onto the abscissa. The confidence interval (x_{l}, x_u) for the true value of the equivalence point is given by the projection on the abscissa of the common surface delimited by the four hyperbolic arcs.

For the first line, we get:

$$y_{01} \pm t_1 s_{y_{01}} \tag{26}$$

and for the second line:

$$y_{02} \pm t_2 s_{y_{02}} \tag{27}$$

t_1 and t_2 are the corresponding t Student values for $\alpha/2 = 0{,}05$ and $N_1 - 2$ and $N_2 - 2$ degrees of freedom, respectively. Hence, the lower value x_l of the confidence interval is obtained by solving the following equation:

$$y_{01} - t_1 s_{y_{01}} = y_{02} + t_2 s_{y_{02}} \tag{28}$$

The higher value x_u is obtained from the equation:

$$y_{01} + t_1 s_{y_{01}} = y_{02} - t_2 s_{y_{02}} \tag{29}$$

From Eqs. (1) and (3) we get

$$\hat{y}_1 = \overline{y}_1 + b_1(x - \overline{x}_1) \tag{30}$$

and then the variance of the fitted y_1 value will be given by

$$V[\hat{y}_1] = V[\overline{y}_1] + (x-\overline{x}_1)^2 V[b_1] = \frac{V[y_1]}{\left(\sum w\right)_1} + (x-\overline{x}_1)^2 \frac{V[y_1]}{(S_{XX})_1}$$
$$= \left(\frac{1}{(\sum w)_1} + \frac{(x-\overline{x}_1)^2}{(S_{XX})_1}\right) V[y_1] \tag{31}$$

Note that the variance of the (weighted) mean of the values

$$V[\overline{y}_1] = V\left[\frac{\left(\sum w\right)_1 y_1}{\left(\sum w\right)_1}\right] = V\left[\frac{\left(\sum \left((\sqrt{w})_1 (\sqrt{w})_1\right) y_1\right)}{\left(\sum w\right)_1}\right]$$
$$= \frac{\left(\sum w\right)_1 V[(\sqrt{w})_1 y_1]}{\left(\left(\sum w\right)_1\right)^2} = \frac{V[y_1]}{\left(\sum w\right)_1} \tag{32}$$

and that the mean y_1 value and the slope b_1 are uncorrelated random variables (property, which was also applied in Eq. (12) without further demonstration) as shown as follows. Taking into account that

$\overline{y}_1 = \sum b_1 (\sqrt{w})_1 y_1$ and $b_1 = \sum c_1 (\sqrt{w})_1 y_1$ where

$$b_1 = \frac{(\sqrt{w})_1}{(\sum w)_1} \qquad c_1 = \frac{(\sqrt{w})_1 (x-\overline{x}_1)}{(S_{XX})_1} \tag{33}$$

and then

$$Cov(b_1, c_1) = \left(\sum a_1 c_1\right) V[(\sqrt{w})_1 y_1] = \left(\sum \frac{(\sqrt{w})_1 (x-\overline{x}_1)}{(S_{XX})_1 \left(\sum w\right)_1}\right) V[y] = 0 \tag{34}$$

From Eq. (31), we get for the standard error of the fitted value

$$s_{y01} = V[\hat{y}_{1i}] = \left(\sqrt{\frac{1}{\left(\sum w\right)_1} + \frac{(x-\overline{x}_1)^2}{(S_{XX})_1}}\right) s_1 \left(y_{01} = \hat{y}_{1i}; \quad s_1 = \sqrt{V[y_1]}\right) \tag{35}$$

Thus, the lower value, x_l, for the confidence interval is obtained by solving the equation

$$a_1 + b_1 x_l + t_1 s_1 \sqrt{\frac{1}{\left(\sum w\right)_1} + \frac{(x_l-\overline{x}_1)^2}{(S_{XX})_1}} = a_2 + b_2 x_l + t_2 s_2 \sqrt{\frac{1}{\left(\sum w\right)_2} + \frac{(x_l-\overline{x}_2)^2}{(S_{XX})_2}} \tag{36}$$

by, for example, successive approximations with an Excel spreadsheet. The higher value x_u is obtained in the same way from the equation that follows also by successive approximations

$$a_1 + b_1 x_u + t_1 s_1 \sqrt{\frac{1}{\left(\sum w\right)_1} + \frac{(x_u - \overline{x}_1)^2}{(S_{XX})_1}} = a_2 + b_2 x_u + t_2 s_2 \sqrt{\frac{1}{\left(\sum w\right)_2} + \frac{(x_u - \overline{x}_2)^2}{(S_{XX})_2}} \tag{37}$$

The point of view of Liteanu et al. [57, 58] is very interesting: the authors consider the point of intersection x_I as belonging to the linear regression before the equivalence point. Then, a certain interval is associated with it. If it is regarded as belonging to the linear regression after the equivalence point, however, another interval is associated with it. As the equivalence point belongs concurrently to both linear regressions, the confidence interval of the two segments can be got by taking the weighted averages of the branches of the two separate sets of confidence intervals. So, we obtain the ultimate confidence interval (x_I^l, x_I^u) where

$$\frac{(N_1 - 2)(x_I)_1^l + (N_2 - 2)(x_I)_2^l}{N_1 + N_2 - 4} = x_I^l \tag{38}$$

$$\frac{(N_1 - 2)(x_I)_1^u + (N_2 - 2)(x_I)_2^u}{N_1 + N_2 - 4} = x_I^u \tag{39}$$

The two values of the limits of the confidence interval will be given by the two solutions of the equations

$$y_I = a_1 + b_1 (x_I)_1 \pm t_1 s_1 \sqrt{\frac{1}{\left(\sum w\right)_1} + \frac{\left((x_I)_1 - \overline{x}_1\right)^2}{(S_{XX})_1}} \tag{40}$$

$$y_I - a_2 + b_2 (x_I)_2 \pm t_2 s_2 \sqrt{\frac{1}{\left(\sum w\right)_2} + \frac{\left((x_I)_2 - \overline{x}_2\right)^2}{(S_{XX})_2}} \tag{41}$$

As the estimation method used assumes the worst case in combining random error of the two lines, the derived confidence limits are on the pessimistic (i.e., realistic) side.

On rearrangement Eq. (40) and squaring, we have

$$\left(y_I - a_1 - b_1 (x_I)_1\right)^2 = t_1^2 s_1^2 \left(\frac{1}{\left(\sum w\right)_1} + \frac{\left((x_I)_1 - \overline{x}_1\right)^2}{(S_{XX})_1}\right) \tag{42}$$

which by simple algebra it may be ordered in powers of x_1 as

$$\begin{aligned} &\left(b_1^2 - \frac{t_1^2 s_1^2}{(S_{XX})_1}\right)(x_I)_1^2 - 2\left(b_1 (y_I - a_1) - t^2 \frac{\overline{x}_1 s_1^2}{(S_{XX})_1}\right)(x_I)_I \\ &+ (y_I - a_1^2) - t_1^2 s_1^2 \left(\frac{1}{\left(\sum w\right)_1} + \frac{\overline{x}^2}{(S_{XX})_1}\right) = 0 \end{aligned} \tag{43}$$

and taking into account the values of $V[b_1]$, $Cov[a_1, b_1]$ and $V[a_1]$ (see **Table 3**), we get finally

$$(b_1^2 - t_1^2 V[b_1])(x_I)_1^2 - 2\Big(b_1(y_I - a_1) + t_1^2 Cov(a_1, b_1)\Big)(x_I)_1 + (y_I - a_1)^2 - t_1^2 V[a_1] = 0 \tag{44}$$

whose roots give the two values of $(x_I)_1$. Since the point of intersection x_I belongs to one of the response functions, then a certain confidence interval is associated with it.

Similarly, if it is regarded as belonging to the other response function, there is another confidence interval associated with it

$$(b_2^2 - t_2^2 V[b_2])(x_I)_2^2 - 2\Big(b_2(y_I - a_2) + t_2^2 Cov(a_2, b_2)\Big)(x_I)_2 + (y_I - a_2)^2 - t_2^2 V[a_2] = 0 \tag{45}$$

Because the intersection point belongs concomitantly to the two response functions, the two segments which together compose the confidence interval, will be obtained by averaging the segments of the two separate confidence intervals, Eqs. (40) and (41). The two values of the limits of the confidence interval will be the two solutions of the second degree Eqs. (44) and (45).

The bands mentioned in this section are [63] for the ordinate of the true line at only a single point. If we desire the confidence bands for the entire line, the critical constant $\sqrt{2F^{\alpha}_{2,n-2}}$ should be substituted for $t_{\alpha/2}$, originating wider bands.

7. Statistical uncertainty of endpoint differences

When we are dealing with the titration of a mixture of a strong and a weak acid that is, hydrochloric and acetic acids, then if x_I is the volume at which the straight lines one and two intersect and x_{II} the volume at which the two and three lines intersect, the difference x_{II}–x_I denoted as Δx, is given by

$$\Delta x = \hat{x}_{II} - \hat{x}_I = \frac{-\Delta a_2}{\Delta b_2} + \frac{\Delta a_1}{\Delta b_1} \tag{46}$$

By multiplying Δx by the molarity of titrant, we have the amount in millimoles of the second acid, that is, acetic acid, in the reaction mixture.

First-order propagation of variance applied to Δx leads to [65] the following expression

$$V[\Delta x] = V[\hat{x}_I] + V[\hat{x}_{II}] + Cov(\Delta a_1, \Delta a_2) + Cov(\Delta a_1, \Delta b_2) + Cov(\Delta a_2, \Delta b_1) + Cov(\Delta b_1, \Delta b_2) \tag{47}$$

where

$$Cov(\Delta a_1, \Delta a_2) = 2\left(\frac{V[\overline{y}_2] + \overline{x}_2^2 V[b_2]}{\Delta b_1 \Delta b_2}\right) \tag{48}$$

$$Cov(\Delta a_1, \Delta b_2) = -2\left(\frac{\overline{x}_2 \hat{x}_{II} V[b_2]}{\Delta b_1 \Delta b_2}\right) \tag{49}$$

$$Cov(\Delta a_2, \Delta b_1) = -2\left(\frac{\overline{x}_2 \hat{x}_I V[b_2]}{\Delta b_1 \Delta b_2}\right) \tag{50}$$

$$Cov(\Delta b_1, \Delta b_2) = 2\left(\frac{\hat{x}_I \hat{x}_{II} V[b_2]}{\Delta b_1 \Delta b_2}\right) \tag{51}$$

The standard error estimate is given by

$$s_{\Delta x} = \sqrt{V[\hat{x}_I] + V[\hat{x}_{II}] + \sum_4 Cov} \tag{52}$$

where $\sum_4$ is the sum of Eqs. (48)–(51).

Attempts to derive confidence limits for Δx as we get in the previous Fieller's theorem section fails because the quantity analogous to z of Eq. (20) involves products of random variables. Therefore, this quantity is not normally distributed and so exact confidence limits cannot be found in terms of Student's t distribution. Because in this case the exact confidence limits cannot be calculated, we use the small variance confidence interval

$$C.I. = 2t_{\alpha/2} s_{\Delta x} \tag{53}$$

8. Application to experimental system

A bibliographic search allows us to demonstrate the importance of conductivity measurements despite their antiquity. The general fundamentals of this technique are collected in Gelhaus and Lacourse (2005) [74] and Gzybkoski (2002) [75]. Its importance in the educational literature has been highlighted [76, 77] and many examples have been recently published in the Journal of Chemical Education i.e., studies on sulfate determination [78]; the identification and quantification of an unknown acid [79], electrolyte polymers [80, 81], acid and basic constants determinations [82], its use in general chemistry [83], microcomputer interface [84] and conductometric-potentiometric titrations [85]. An accurate method of determining conductivity in acid-base reactions [86], the acid-base properties of weak electrolytes [87], and those of polybasic organic acids [88] have also been recently subject of study.

The relation between conductometric and the volume of titrant added leads to segmented linear titration curves, the endpoint being defined by the intersection of the two straight lines segments. What follows is the application of the possible methods of uncertainty estimation of the endpoint of data described in the literature as well as experimental measurements carried out in the laboratory.

8.1. Conductometric titration of 100 mL of a mixture of acids with potassium hydroxide 0.100 M

Table 5 shows the data [conductance (1/*R*), volume (*x*)] published by Carter et al. [69]; Schwartz and Gelb [65]) and corresponding to the conductometric titration of a mixture of acids, perchloric acid and acetic acid with potassium hydroxide 0.100 M as titrant agent. The points recorded belong to the three branches of the titration curve; the first (branch A) corresponds to the neutralization of perchloric acid, the second (branch B) to the neutralization of acetic acid, and the third (branch C) to the excess of potassium hydroxide.

Let us focus first on the perchloric acid titration. The plot of conductance data (1/*R*) versus volume (*x*), in general, is not linear due to the dilution effect of the titrant. So that, as it is carried out in the usual way, it is plotted the product $(1/R)(100 + x)$ versus x (see **Figure 1**).

Firstly, Schwartz and Gelb [65] select 13 points, six (volume 4–14 mL) for branch A and seven (volume 20–32 mL) for branch B. The points near to the endpoint of perchloric acid are deviating from linearity and discarded in the first instance. It is also considered that the data have a different variance $V{\cdot}(100 + x_i)^2$, being the weighting factor $(100 + x_i)^{-2}$ (see **Table 5**).

In the case of acetic titration, six points (volume 35–44 mL) are selected for branch C, at first. The points of branch B near to the acetic acid endpoint are discarded. **Figures 2** and **3** show the straight line segments with the corresponding selected points.

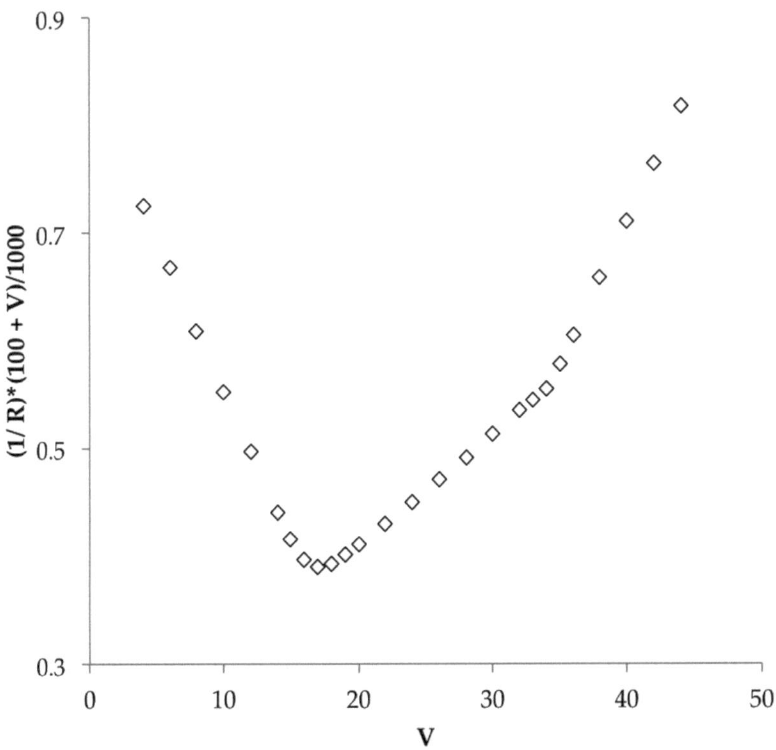

Figure 1. Conductometric titration of a mixture of perchloric and acetic acids with potassium hydroxide (data shown in **Table 5**).

1/*R*	*x*	*y*	W_i	1/*R*	*x*	*y*	W_i
6.975	4	0.7254	9.246E−05	3.633	24	0.4505	6.504E−05
6.305	6	0.6683	8.900E−05	3.742	26	0.4715	6.299E−05
5.638	8	0.6089	8.573E−05	3.840	28	0.4915	6.104E−05
5.020	10	0.5522	8.264E−05	3.946	30	0.5130	5.917E−05
4.432	12	0.4964	7.972E−05	4.052	32	0.5349	5.739E−05
3.865	14	0.4406	7.695E−05	4.097	33	0.5449	5.653E−05
3.610	15	0.4152	7.561E−05	4.145	34	0.5554	5.569E−05
3.415	16	0.3961	7.432E−05	4.280	35	0.5778	5.487E−05
3.328	17	0.3894	7.305E−05	4.445	36	0.6045	5.407E−05
3.330	18	0.3929	7.182E−05	4.772	38	0.6585	5.251E−05
3.370	19	0.4010	7.062E−05	5.080	40	0.7112	5.102E−05
3.420	20	0.4104	6.944E−05	5.380	42	0.7640	4.959E−05
3.522	22	0.4297	6.719E−05	5.680	44	0.8179	4.823E−05

Table 5. Data conductance (1/*R*) and volume (*x*) corresponding to the titration of a mixture of perchloric acid and acetic acid with potassium hydroxide.

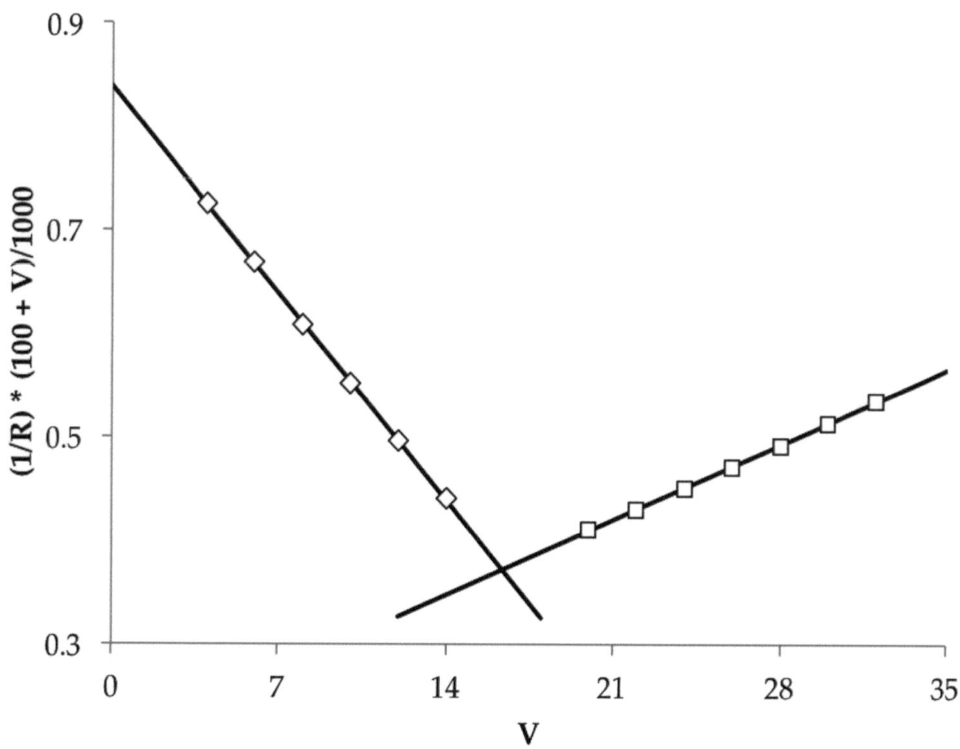

Figure 2. Conductometric titration of perchloric acid in the mixture (branches A and B).

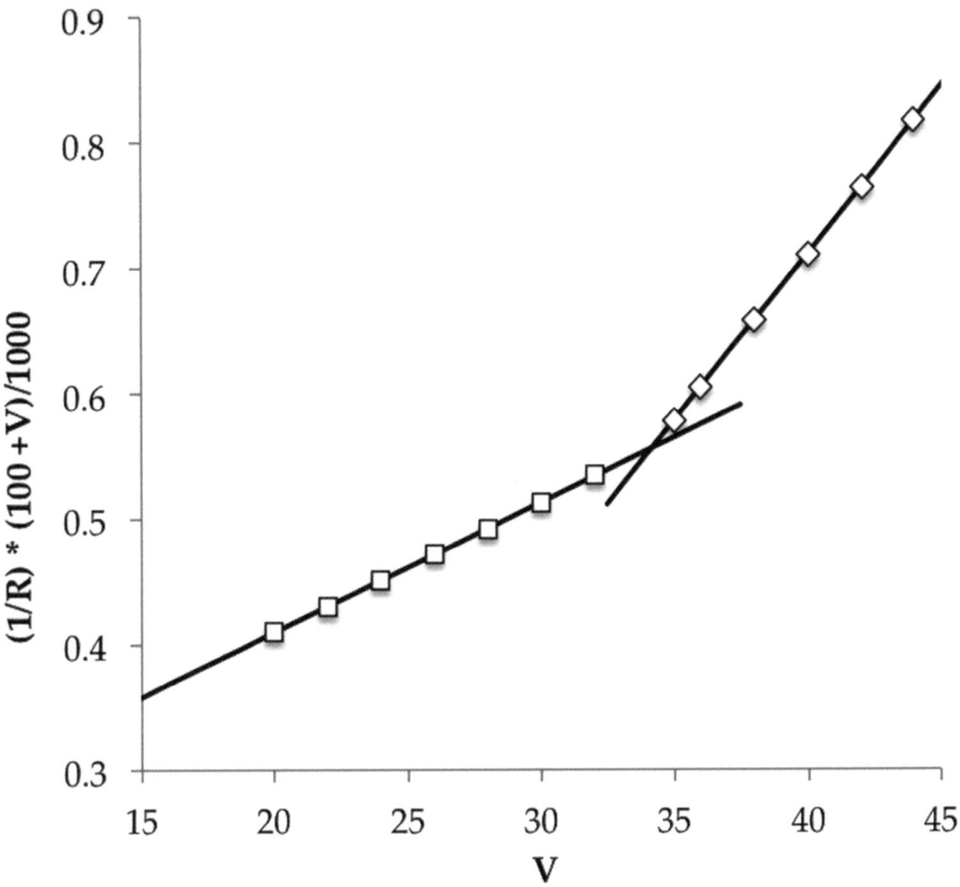

Figure 3. Conductometric titration of acetic acid in the mixture (branches B and C).

Table 6 includes the intermediate results obtained in the calculation of the first endpoint, corresponding to the neutralization of perchloric acid (**Figure 2**), in order to follow the procedures previously detailed. The first endpoint is located at 16.367 mL and therefore 1.637 mmol of $HClO_4$. The estimated standard error at the endpoint, using the first-order propagation of variance, is 0.039 mL. The confidence limits are calculated using $t = 2.262$ (9 degrees of freedom) and correspond to 16.455 and 16.279 mL, respectively, for the upper and lower limits, being the confidence interval equal to 0.176 mL. The application of Fieller's theorem leads to the values of 16.455 and 16.278 mL, respectively. Carter et al. [67] give values of 16.455 and 16.279 mL, identical to the first ones indicated.

The second endpoint, corresponding to the complete neutralization of both perchloric and acetic acids, is located at 34.197 mL. If $x_{(I)}$ is the volume in which lines A and B intersect, and $x_{(II)}$ the volume in which lines B and C intersect, the difference $x_{(II)}$–$x_{(I)}$ (34.1971–16.3665 mL) corresponds to acetic acid in the sample, 17.831 mL. If the above methodology is used for lines, B and C (**Figure 3**) give $x_{(II)} \pm s_{d\,[x(II)]}$ equal to 34.197 ± 0.0478, and 34.305 and 34.089 mL for the confidence limits.

http://dx.doi.org/10.5772/intechopen.68827

1/R	x	y = (1/R)(100 + x)	$(100 + x_i)^{-2}$			1/R	x	y = (1/R)(100 + x)	$(100 + x_i)^{-2}$
6.975	4	0.7254	9.246E−05			3.420	20	0.4104	6.944E−05
6.305	6	0.6683	8.900E−05			3.522	22	0.4297	6.719E−05
5.638	8	0.6089	8.573E−05			3.633	24	0.4505	6.504E−05
5.020	10	0.5522	8.264E−05			3.742	26	0.4715	6.299E−05
4.432	12	0.4964	7.972E−05			3.840	28	0.4915	6.104E−05
3.865	14	0.4406	7.695E−05			3.946	30	0.5130	5.917E−05
						4.052	32	0.5349	5.739E−05
N_1=	6	$[\Sigma W_i]1$=	5.065E−04			N_2=	7	$[\Sigma W_i]2$=	4.423E−04
x_1(mean)=	8.786	y_1(media)=	0.5881			x_2(mean)=	25.746	y_2(mean)=	0.4690
		$[S(xx)]1$=	5.8987E−03					$[S(xx)]2$=	7.0636E−03
		$[S(xy)]1$=	−1.683E−04					$[S(xy)]2$=	7.323E−05
		$[S(yy)]1$=	4.804E−06					$[S(yy)]2$=	7.594E−07
	b_1=	−0.028535824		Δb=	0.038903		b_2=	0.010367102	
	a_1=	0.838792976		Δa=	−0.636705		a_2=	0.202087987	
	$[R2]1$=	0.99988		$x(I)$=	16.3665		$[R2]2$=	0.99974	
	$V[y/x]1$=	1.45580E−10					$V[y/x]2$=	3.96396E−11	
	$V(b_1)$=	2.468E−08		V[pooled]=	8.67244E−11		$V(b_2)$=	5.612E−09	
	$V(a_1)$=	2.193E−06					$V(a_2)$=	3.809E−06	
	$Cov(a_1, b1)$=	−2.168E−07					$Cov(a_2, b_2)$=	−1.445E−07	
$V[\Delta a]$=	6.002E−06	$V[\Delta a]$=	9.640E−06	$V[x(I)]$=	0.0015	FIELLER	$ax2 + bx + c = 0$		
$V[\Delta b]$=	3.029E−08	$V[\Delta b]$=	2.698E−08	$s[x(I)]$=	0.0389	a=	1.513E−03	$V(u)$=	16.455
$cov(\Delta a, \Delta b)$=	−3.613E−07	$cov(\Delta a, \Delta b)$=	−4.453E−07	$t(0, 05, 9)$=	2.262	b=	−4.953E−02	$V(I)$=	16.278
		Pooled variances		$t\cdot s[x(I)]$=	0.0880	c=	4.053E−01		

Table 6. Intermediate results obtained in the calculation of the first endpoint (titration of perchloric acid with potassium hydroxide (**Figure 2**)).

However, as it is indicated in the section on "statistical uncertainty of endpoint differences," the statistical uncertainty of Δx is not a simple combination of uncertainties for $x_{(I)}$ and $x_{(II)}$. The attempt to deduce equations analogous to Eqs. (22) and (25) in order to calculate the confidence limits for Δx, is not applicable since the magnitude analogous to z in Eq. (20) implies, in this case, the product of random variables.

This quantity is not normally distributed, and therefore, no exact confidence limits can be calculated in terms of the Student t distribution. The application of (first-order) propagation of the variance is nonetheless feasible, leading this procedure to an expression for the standard error of Δx of the same type as Eq. (9) for a single endpoint.

The latter methodology is applied to the optimal case detailed by Schwartz and Gelb [65]. The corresponding data are shown in **Figure 4**, and the calculations necessary to locate the equivalence points, first and second, are shown in **Table 7**. The results obtained are: first equivalence point (perchloric acid): $x_{(I)} = 16.358$ mL, $s[x_{(I)}] = 0.035$ mL, $t\,s\,[x_{(I)}] = 0.078$ mL, $[\text{I.C.}]_I = 0.156$ mL. Second equivalence point (mixture of perchloric and acetic acids): $x_{(II)} = 34.244$ mL, $s[x_{(II)}] = 0.027$ mL, $t\,s\,[x_{(II)}] = 0.061$ mL, $[\text{IC}]_{II} = 0.122$ mL. This latter is not correct because it does not take into account the covariances described in Section 7. If covariances are incorporated into the calculations, we get for the second point (acetic acid): $x = 17.887$ mL, $s[\Delta x] = 0.040$ mL; $t\,s[\Delta x] = 0.086$ mL, $[\text{IC}]_{\Delta x} = 0.172$ mL. The confidence interval, as expected, is higher than that found for $x_{(II)}$, despite decreasing the value of Student's t by increasing the number of degrees of freedom: $N_1 + N_2 - N_3 - 2 \times 3 = 13$).

Some points near to the endpoint appear to deviate slightly from linearity. However, it is not always clear whether or not to omit these problem points in the analysis, which can be done by

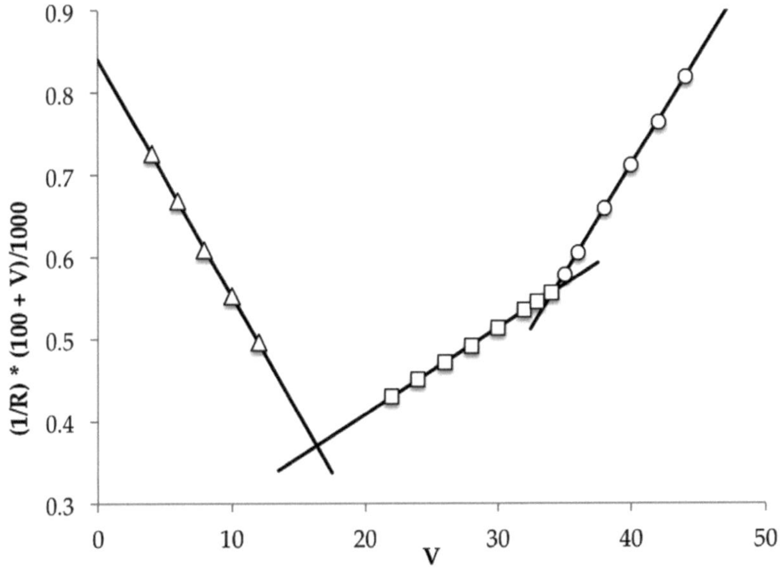

Figure 4. Illustrative example described by Schwartz and Gelb [65] as optimal. Numerical data are shown in **Table 5**. First branch (A), volumes of 4–12 mL, 5 points. Second branch (B), volumes of 22–34 mL, 8 points. Third branch (C), volumes of 35–44 mL, 6 points.

1/*R*	***x***	***y* = (1/*R*)(100 + *x*)**	**$(100 + x_i)-^2$**			**1/*R***	***x***	***y* = (1/*R*)(100 + *x*)**	**$(100 + x_i)-^2$**
6.975	4	0.7254	9.246E−05			3.522	22	0.4297	6.719E−05
6.305	6	0.6683	8.900E−05			3.633	24	0.4505	6.504E−05
5.638	8	0.6089	8.573E−05			3.742	26	0.4715	6.299E−05
5.020	10	0.5522	8.264E−05			3.840	28	0.4915	6.104E−05
4.432	12	0.4964	7.972E−05			3.946	30	0.5130	5.917E−05
						4.052	32	0.5349	5.739E−05
						4.097	33	0.5449	5.653E−05
						4.145	34	0.5554	5.569E−05
N_1=	5	$[\Sigma W_i]1$=	4.296E−04			N_2=	8	$[\Sigma W_i]2$=	4.850E−04
x_1(mean)=	7.852	y_1(media)=	0.6145			x_2(mean)=	28.362	y_2(mean)=	0.4962
		$[S(xx)]1$=	3.4319E−03					$[S(xx)]2$=	8.221E−03
		$[S(xy)]1$=	−9.856E−05					$[S(xy)]2$=	8.624E−05
		$[S(yy)]1$=	2.831E−06					$[S(yy)]2$=	9.048E−07
	b_1=	−0.028717584		Δb=	0.039208		$b2$=	0.010490769	
	$a1$=	0.839983924		Δa=	−0.641362		$a2$=	0.198621917	
	$[R2]1$=	0.99989		$x(I)$=	16.3578		$[R2]2$=	0.99992	
	$V[y/x]1$=	1.03732E−10					$V[y/x]2$=	1.25234E−11	
	$V(b_1)$=	3.023E−08		V[pooled]=	4.29264E−11		$V(b_2)$=	1.523E−09	
	$V(a_1)$=	2.105E−06					$V(a_2)$=	1.251E−06	
	$cov(a_1, b_1)$=	-2.373E−07					$cov(a_2, b_2)$=	-4.321E−08	
$V[\Delta a]$=	3.356E−06	$V[\Delta a]$=	5.160E−06	$V[x(I)]$=	0.0012	FIELLER	$ax2 + bx + c = 0$		
$V[\Delta b]$=	3.175E−08	$V[\Delta b]$=	1.773E−08	$s[x(I)]$=	0.0347	a=	1.537E−03	$V(u)$=	16.436
$cov(\Delta a, \Delta b)$=	−2.805E−07	$cov(\Delta a, \Delta b)$=	−2.463E−07	$t(0, 05, 9)$=	2.262	b=	−5.029E−02	$V(l)$=	16.280
		Pooled variances		$t\ s[x(I)]$=	0.0784	c=	4.113E−01		
		$V(u)$=	16.436	$V(l)$=	16.279				

1/R	x	y = (1/R)(100 + x)	$(100 + x_i)^{-2}$			1/R	x	y = (1/R)(100 + x)	$(100 + x_i)^{-2}$
4.280	35	0.5778	5.487E−05		Δb=	0.016147			
4.445	36	0.6045	5.407E−05		Δa=	−0.552936			
4.772	38	0.6585	5.251E−05		$x(I)$=	34.2444			
5.080	40	0.7112	5.102E−05		Δx=	17.8866			
5.380	42	0.7640	4.959E−05						
5.680	44	0.8179	4.823E−05		V[pooled]=	1.12632E−11			
N_3=	6	$[\Sigma W_i]1$=	3.103E−04						
x_3(mean)=	39.022	y_3(media)=	0.6851						
		$[S(xx)]1$=	3.1199E−03						
		$[S(xy)]1$=	8.311E−05						
		$[S(yy)]1$=	2.214E−06						
				$V[x(I)]$=	0.0007	$t(0, 05, 9)$=	2.262	$V(u)$=	34.305
	$b3$=	0.026637531		$s[x(I)]$=	0.0269	$t\ s[x(I)]$=	0.0610	$V(l)$=	34.183
	$a3$=	−0.354313932							
	$[R2]3$=	0.99998							
	$V[y/x]3$=	9.37285E−12		$V[x(I)]$=	1.201E−03				
	$V(b3)$=	3.004E−09		$V[x(II)]$=	7.262E−04				
	$V(a3)$=	4.605E−06		$Cov(\Delta a_1, \Delta a_2)$=	3.871E−03				
	$cov(a_3, b_3)$=	−1.172E−07		$Cov(\Delta a_1, \Delta b_2)$=	−4.674E−03				
				$Cov(\Delta a_2, \Delta b_1)$=	−2.233E−03				
$V[\Delta a]$=	5.761E−12	$V[\Delta a]$=	6.659E−06	$Cov(\Delta b_1, \Delta b_2)$=	2.696E−03				
$V[\Delta b]$=	4.528E−09	$V[\Delta b]$=	4.980E−09						
$cov(\Delta a, \Delta b)$=	−1.294E−07	$cov(\Delta a, \Delta b)$=	−1.797E−07	$V[\Delta x]$=	1.587E−03	$t(0, 05, 13)$=	2.160	$V(u)$=	17.973
		Pooled variances		$s[\Delta x]$=	3.984E−02	$t\ s[\Delta x]$=	0.086	$V(l)$=	17.801

Table 7. Evaluation of endpoints in the titration of a mixture of $HClO_4$ and CH_3COOH with KOH 0.100 M, optimum case (**Figure 4**).

trial and error. The optimal point set (**Figure 4**) is one that minimizes, for example, the confidence interval [63].

The weighting factors are very similar so that the values obtained by weighted linear regression and the simple one become equivalent.

8.2. Conductometric titration of hydrochloric acid 0.1 M with sodium hydroxide 0.1 M

The data corresponding to the two branches of the conductometric titration of 0.1 M HCl with 0.1 M NaOH is shown in the upper part of **Table 8** and plot in **Figure 5**. The cut-off point of both lines is (6.414, 0.358) [57, 58, 89].

Table 8 also shows all the operations required to calculate the minimum and maximum values of the confidence interval by the use of hyperbolic confidence bands for the two linear branches. The limit x_l of the confidence interval is obtained by solving Eq. (36), which in this case (**Table 8**) is

$$\begin{aligned}\Theta_l &= 1.403 - 0.0637x_l + 1.943 \cdot 0.01034\sqrt{\frac{1}{8} + \frac{(x_l - 9)^2}{168}} \\ &+0.4908 - 0.0517x_l + 2.353 \cdot 0.0024\sqrt{\frac{1}{5} + \frac{(x_l - 20.2)^2}{32.8}} = 0\end{aligned} \tag{54}$$

leading to xl = 16.264 mL. The highest value is obtained by solving (Eq. (37))

$$\begin{aligned}\Theta_u &= 1.403 - 0.0637x_u + 1.943 \cdot 0.01034\sqrt{\frac{1}{8} + \frac{(x_u - 9)^2}{168}} \\ &+0.4908 - 0.0517x_u + 2.353 \cdot 0.0024\sqrt{\frac{1}{5} + \frac{(x_u - 20.2)^2}{32.8}} = 0\end{aligned} \tag{55}$$

which leads to x_u = 16.564 mL. Both equations θ_l = 0 and θ_u = 0 are resolved by successive approximations. Different values are tested for the lower and upper limits to get a change of sign in θ_l and θ_u.

In the weighted mean method (**Table 8**), the following equations are solved

$$0.00405(x_I)_1^2 - 0.1332(x_I)_1 + 1.09179 = 0 \tag{56}$$

$$0.00267(x_I)_2^2 - 0.08776(x_I)_2 + 0.720 = 0 \tag{57}$$

being resulted from squaring and reordering the Eqs. (44) and (45), respectively (expressed as a function of the variances of a_1, b_1 and of the covariance between a_1 and b_1). Once calculated the solutions of the Eqs. (56): 16.630 and 16.487 mL, and (57): 16.206 and 16.339 mL, we have

	x	y			x	y	
	2	1.265			17	0.388	
	4	1.141			18	0.441	
	6	1.028			20	0.544	
	8	0.906			22	0.644	
	10	0.777			24	0.752	
	12	0.641					
	14	0.51					
	16	0.372					
$N1$=	8			$N2$=	5		
MEAN1=	9	0.83		MEAN2=	20.2	0.5538	
[SXX]1=	168			[SXX]2=	32.8		
a_1=	-0.06367	1.40300	$=a_0$	a_1=	0.05171	-0.49081	$=a_0$
$s(a_1)$=	0.00080	0.00806	$=s(a_0)$	$s(a_1)$=	0.00036	0.00726	$=s(a_0)$
$R2$=	0.99906	0.01034	$=s(y/x)$	$R2$=	0.99986	0.00204	$=s(y/x)$
		$x(I)$=	16.414	$y(I)$=	0.358		
$t(0.05;6)$=	1.943	$t_1\ s(y/x)1$=	0.0201	$t(0.05;3)$=	2.353	$t_2\ s(y/x)2$=	0.0048
	θ	DIFF-1	DIFF-2	θ	DIFF-1	DIFF-2	
	16.25	0.00164	0.036126	16.50	−0.02736	0.007436	
	16.26	0.00048	0.034979	16.51	−0.02852	0.006289	
	16.261	0.00037	0.034864	16.52	V0.02968	0.005141	
	16.262	0.00025	0.034749	16.53	−0.03084	0.003994	
	16.263	0.00014	0.034634	16.54	−0.03200	0.002846	
	16.264	0.0000191	0.034520	16.55	−0.03316	0.001699	
	16.2641	0.0000075	0.034508	16.56	−0.03432	0.000552	
	16.2642	−0.0000041	0.034497	16.561	−0.03443	0.000437	
	16.265	−0.00010	0.034405	16.562	−0.03455	0.000322	
	16.266	−0.00021	0.034290	16.563	−0.03467	0.000207	
	16.267	−0.00033	0.034175	16.564	−0.03478	0.000093	
	16.268	−0.00044	0.034060	16.5648	−0.03487	0.000001	
	16.269	−0.00056	0.033946	16.5649	−0.03489	−0.000011	
	16.27	−0.00068	0.033831	16.565	−0.03490	−0.000022	
			16.265	16.414	16.565		

Table 8. Hyperbolic confidence intervals for the two lines: successive approximations.

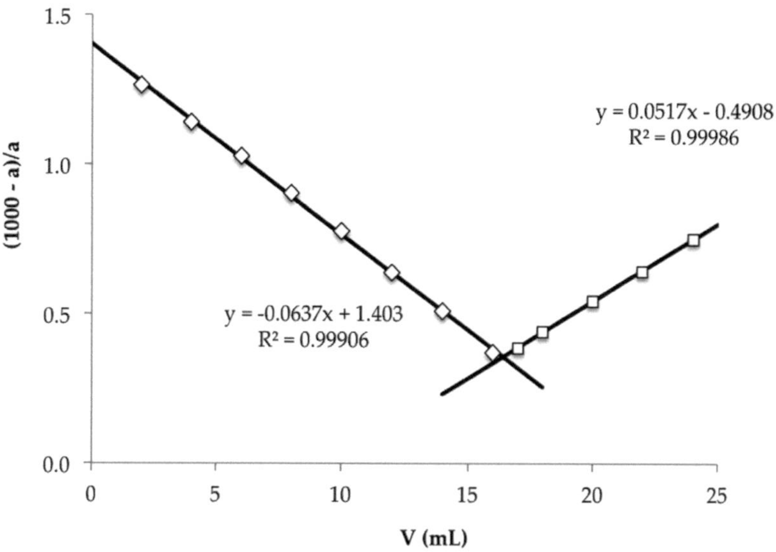

Figure 5. Conductometric titration of hydrochloric acid 0.1 M with sodium hydroxide 0.1 M as a titrant (data are shown in **Table 8**).

$$x_u = \frac{(8-2)16.630 + (5-2)16.487}{8+5-4} = 16.583\ mL \tag{58}$$

$$x_l = \frac{(8-2)16.206 + (5-2)16.339}{8+5-4} = 16.250\ mL \tag{59}$$

8.3. Experimental measurements: conductometric titration of 100 mL of a mixture of hydrochloric acid and acetic acids with potassium hydroxide 0.100 M

8.3.1. Reagents

Acetic acid ($C_2H_4O_2$) $M = 60$ g/mol (MERCK > 99.5%; 1.049 g/mL); hydrochloric acid (HCl) 1 M (MERCK, analytical grade); potassium hydroxide (KOH) 1 M (MERCK, analytical grade); potassium hydrogen phthalate ($C_8H_5KO_4$) $M = 204.23$ g/mol (MERCK > 99.5%).

8.3.2. Instruments

4-decimal point analytical balance (Metler AE200), conductivity meter Crimson (EC-Metro GLP 31), calibrated by standards of 147 μS/cm, 1413 μS/cm, 12.88 mS/cm. Digital burette of 50 mL (Brand) (accuracy: 0.2%, precision: <0.1%, resolution: 0.01 mL, with standard vent valve at 20°C).

8.3.3. Solutions

- Mixture of hydrochloric and acetic acids 0.015 M.

- Potassium hydroxide 0.1 M.

8.3.4. Experimental

About 100 mL of mixture of hydrochloric and acetic acids 0.015 M is transferred to a 250 mL volumetric flask containing 100 mL of distilled water. Then, the mixture is titrated conductometrically with KOH 0.0992 ± 0.0001 M ($n = 3$), (previously standardized with potassium hydrogen phthalate). **Table 9** shows the data [conductance, volume] as well as the product of the conductance by $(100 + x)/100$ to correct the dilution effect of the titrant. The data are plotted in **Figure 6.**

V KOH (mL)	Conductance (mS/cm)	Conductance* (mS/cm)	V KOH (mL)	Conductance (mS/cm)	Conductance* (mS/cm)
0.0	5.64	5.6400	21.1	2.22	2.6884
1.3	5.30	5.3689	21.5	2.25	2.7338
2.1	5.10	5.2071	22.0	2.27	2.7694
4.6	4.44	4.6442	23.0	2.34	2.8782
5.5	4.21	4.4416	24.0	2.39	2.9636
6.1	4.05	4.2971	25.0	2.45	3.0625
7.0	3.82	4.0874	26.1	2.51	3.1651
8.0	3.58	3.8664	27.0	2.56	3.2512
9.0	3.32	3.6188	28.0	2.61	3.3408
10.0	3.07	3.3770	29.1	2.67	3.4470
11.0	2.84	3.1524	30.0	2.72	3.5360
12.1	2.58	2.8922	31.0	2.88	3.7728
13.1	2.36	2.6692	32.0	3.03	3.9996
14.0	2.16	2.4624	33.0	3.19	4.2427
15.0	2.01	2.3115	34.0	3.35	4.4890
15.5	1.97	2.2765	35.0	3.50	4.7250
16.0	1.96	2.2736	36.0	3.63	4.9368
16.5	1.97	2.2951	37.0	3.79	5.1923
17.1	1.99	2.3303	38.0	3.91	5.3958
17.5	2.01	2.3618	39.0	4.04	5.6156
18.0	2.04	2.4072	40.0	4.19	5.8660
18.5	2.06	2.4411	41.0	4.32	6.0912
19.0	2.09	2.4871	42.1	4.46	6.3377
19.5	2.12	2.5334	43.0	4.57	6.5351
20.0	2.15	2.5800	44.0	4.70	6.7680
20.5	2.18	2.6269	45.0	4.82	6.9890

* Conductivity·((100 + V)/100).

Table 9. Conductance and KOH volume data corresponding to the titration of a mixture of hydrochloric and acetic acids with potassium hydroxide (first assay).

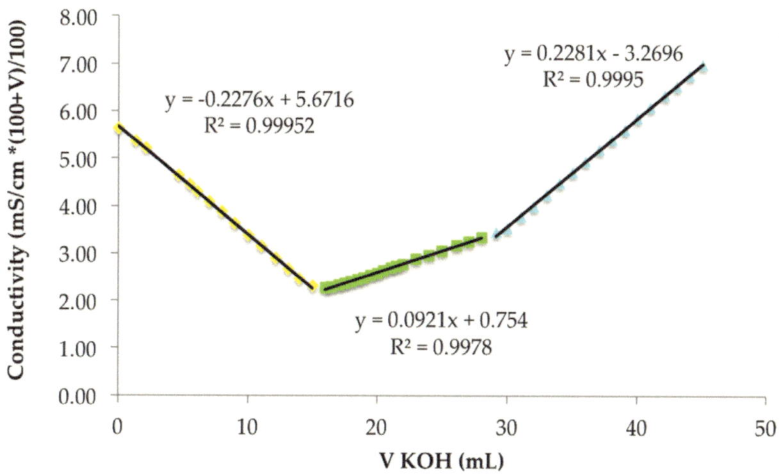

Figure 6. Conductometric titration of a mixture of hydrochloric and acetic acids with potassium hydroxide (data are shown in **Table 9**, first assay). Branch A: V [0–15]. Branch B: V [16–28]. Branch C: V [29.1–45].

The points recorded belong to the three branches of the titration curve; the first (branch A) corresponds to the neutralization of hydrochloric acid, the second (branch B) to the neutralization of acetic acid, and the third (branch C) to the excess of potassium hydroxide.

Figures 7 (hydrochloric acid) and **8** (hydrochloric acid + acetic acid) are the graphs corresponding to the estimation of the endpoints. The points represented in the graph and then

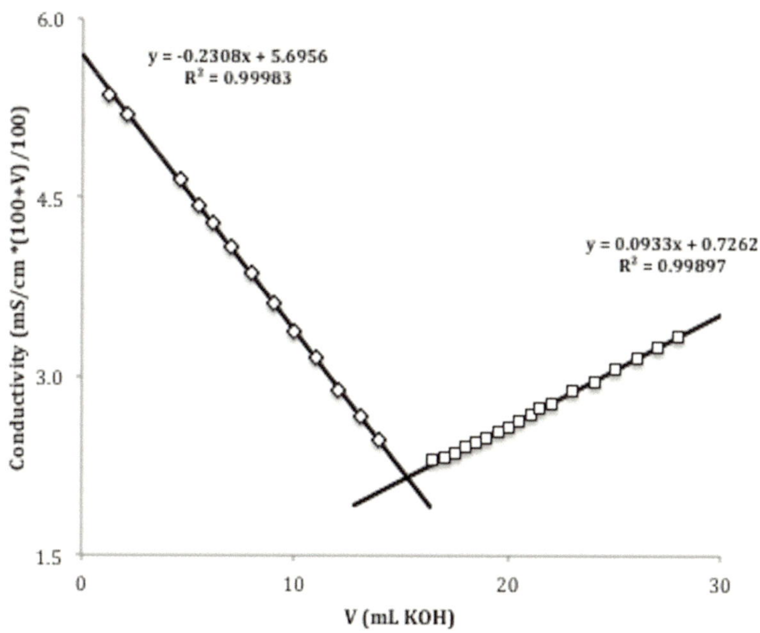

Figure 7. Conductometric titration of hydrochloric acid in the mixture (branches A and B).

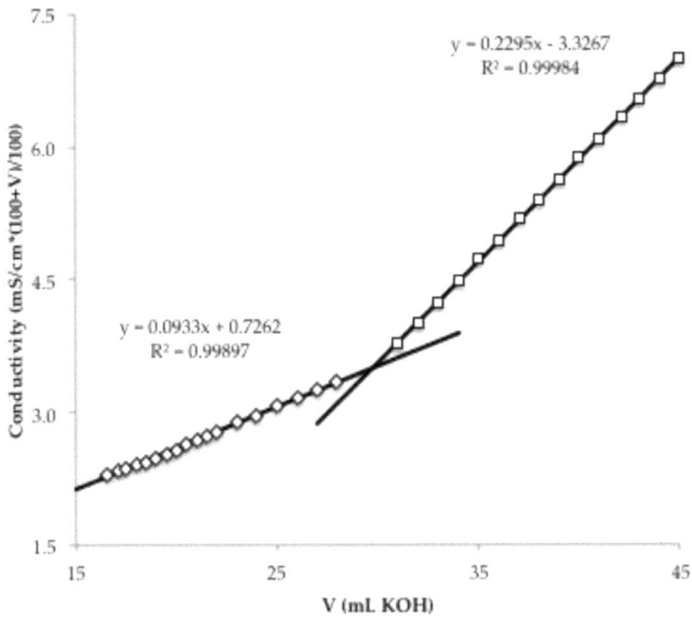

Figure 8. Conductometric titration of acetic acid in the mixture (branches B and C).

used in the calculations are colored yellow (branch A), green (branch B) and blue (branch C) in **Table 9**, thus avoiding proximity to the breakpoints. The values obtained for the intersections of the abscissa are 15.334 mL for hydrochloric acid and 29.743 mL for the sum of hydrochloric and acetic acids. So, acetic acid corresponds to the difference, 14.410 mL. From the data of **Figure 6**, without discarding of points, somewhat different values are obtained: 15.383, 29.582 and 14.189 mL.

Table 10 shows in detail all the calculations necessary to estimate the confidence limits of the abscissa of the breakpoint. The first-order variance propagation method [60] leads to the following volumes ± confidence limits: 15.334 ± 0.0619 (first endpoint), 29.743 ± 0.151 (second endpoint), and 14.410 ± 0.142 (difference). In the second case, the confidence limits cannot refer to the difference (acetic acid), since the covariates involved are not taken into account (as previously explained in Section 7). Three decimal numbers were considered to compare and check calculations.

The application of Fieller's theorem leads to the same results as those obtained by the law of propagation of errors, not being applicable to the estimation of confidence limits of the difference of volumes. The fundamentals of the first-order variance propagation method and Fieller's theorem are much stronger than those based on the use of hyperbolic confidence bands, which lead to higher confidence intervals and limits (not applied in this case).

The conductometric titration was carried out in triplicate, on different days, obtaining the results included in **Tables 11** and **12**, and also represented in **Figures 9** and **10**. Again, the data

N_1=	13	$[\Sigma W_i]1$=	13			$N2$=	18	$[\Sigma W_i]2$=	18
x_1(mean)=	7.985	y_1(media)=	3.8527			x_2(mean)=	21.350	y_2(media)=	2.7175
		$[S(xx)]1$=	194.9369					$[S(xx)]2$=	209.5250
		$[S(xy)]1$=	−44.9940					$[S(xy)]2$=	19.5426
		$[S(yy)]1$=	10.3870					$[S(yy)]2$=	1.8246
	b_1=	−0.230813226		Δb=	0.324084		b_2=	0.093270863	
	a_1=	5.695614832		Δa=	−4.969405		$a2$=	0.726209860	
	$[R2]1$=	0.99983		$x(I)$=	15.3337		$[R2]2$=	0.99897	
	$V[y/x]1$=	1.60452E−04					$V[y/x]2$=	0.000117906	
	$V(b1)$=	8.231E−07		V[pooled]=	0.000135239		$V(b2)$=	5.627E−07	
	$V(a1)$=	6.482E−05					$V(a2)$=	2.631E−04	
	$cov(a_1, b_1)$=	−6.572E−06					$cov(a_2, b_2)$=	−1.201E−05	
$V[\Delta a]$=	3.279E−04	$V[\Delta a]$=	3.564E−04	$V[x(I)]$=	0.0007	FIELLER	$ax2 + bx + c = 0$		
$V[\Delta b]$=	1.386E−06	$V[\Delta b]$=	1.339E−06	$s[x(I)]$=	0.0274	a=	1.050E−01	$V(u)$=	15.390
$cov(\Delta a, \Delta b)$=	−1.859E−05	$cov(\Delta a, \Delta b)$=	−1.932E−05	$t(0, 05, 27)$=	2.052	b=	−3.221E + 00	$V(l)$=	15.278
		Pooled variances		$t\ s[x(I)]$=	0.0562	c=	2.469E + 01		
		$V(u)$=	15.390	$V(l)$=	15.278				

Table 10. Evaluation of endpoints in the titration of a mixture of HCl and CH_3COOH with KOH 0.0992 M (data **Table 9**).

V KOH (mL)	Conductance (mS/cm)	Conductance* (mS/cm)	V KOH (mL)	Conductance (mS/cm)	Conductance* (mS/cm)
0.0	5.71	5.7100	22.0	2.27	2.7694
1.0	5.47	5.5247	23.0	2.33	2.8659
2.0	5.17	5.2734	24.0	2.39	2.9636
3.0	4.90	5.0470	25.0	2.45	3.0625
4.0	4.61	4.7944	26.0	2.51	3.1626
5.0	4.34	4.5570	27.0	2.57	3.2639
6.0	4.06	4.3036	28.0	2.62	3.3536
7.0	3.79	4.0553	29.0	2.67	3.4443
8.0	3.52	3.8016	30.0	2.78	3.6140
9.0	3.28	3.5752	31.0	2.94	3.8514
10.0	3.02	3.3220	32.0	3.10	4.0920
11.1	2.75	3.0553	33.0	3.27	4.3491
12.0	2.53	2.8336	34.0	3.42	4.5828
13.0	2.29	2.5877	35.0	3.57	4.8195
14.0	2.10	2.3940	36.0	3.73	5.0728
15.0	1.96	2.2540	37.0	3.88	5.3156
15.5	1.94	2.2453	38.0	4.03	5.5614
16.0	1.95	2.2585	39.0	4.17	5.7963
16.5	1.96	2.2869	40.0	4.30	6.0200
17.0	1.99	2.3248	41.0	4.44	6.2604
17.5	2.01	2.3618	42.0	4.57	6.4894
18.0	2.04	2.4072	43.0	4.70	6.7210
19.0	2.10	2.4990	44.0	4.83	6.9552
20.0	2.16	2.5920	45.0	4.95	7.1775
21.0	2.22	2.6862			

* Conductivity·((100 + V)/100).

Table 11. Conductance and KOH volume data corresponding to the titration of a mixture of hydrochloric and acetic acids with potassium hydroxide (second assay).

used in the detailed calculations are colored in the tables. The results obtained (and intermediate calculations) for the second assessment are shown in **Table 13**: 14.913 ± 0.041 (propagation of errors and Fieller), 29.372 ± 0.120 (approximate method of propagation of errors) and 14.458 ± 0.113 (propagation of errors). In the third assessment: 15.032 ± 0.043, 29.414 ± 0.146, and 14.383 ± 0.140 mL.

V KOH (mL)	Conductance (mS/cm)	Conductance* (mS/cm)	V KOH (mL)	Conductance (mS/cm)	Conductance* (mS/cm)
0.0	5.81	5.8100	22.0	2.29	2.7938
1.0	5.51	5.5651	23.0	2.34	2.8782
2.0	5.22	5.3244	24.0	2.40	2.9760
3.0	4.94	5.0882	25.0	2.45	3.0625
4.0	4.66	4.8464	26.0	2.52	3.1752
5.0	4.38	4.5990	27.0	2.57	3.2639
6.0	4.11	4.3566	28.0	2.63	3.3664
7.0	3.85	4.1195	29.0	2.68	3.4572
8.0	3.59	3.8772	30.0	2.75	3.5750
9.0	3.32	3.6188	31.0	2.92	3.8252
10.0	3.07	3.3770	32.0	3.09	4.0788
11.0	2.82	3.1302	33.0	3.26	4.3358
12.0	2.58	2.8896	34.0	3.42	4.5828
13.0	2.33	2.6329	35.0	3.56	4.8060
14.0	2.13	2.4282	36.0	3.71	5.0456
14.5	2.05	2.3473	37.0	3.86	5.2882
15.0	1.99	2.2862	38.0	4.00	5.5200
16.0	1.96	2.2748	39.0	4.14	5.7546
16.5	1.98	2.3020	40.0	4.28	5.9920
17.0	2.00	2.3377	41.0	4.40	6.2040
17.5	2.03	2.3853	42.0	4.54	6.4468
18.0	2.05	2.4190	43.0	4.67	6.6781
19.0	2.11	2.5109	44.0	4.80	6.9120
20.0	2.17	2.6040	45.0	4.92	7.1340
21.0	2.23	2.6983			

* Conductivity·((100 + V)/100).

Table 12. Conductance and KOH volume data corresponding to the titration of a mixture of hydrochloric and acetic acids with potassium hydroxide (third assay).

If the series corresponding to the first equivalence point are analyzed: 15.334, 14.913 and 15.032, one of the data seems to be very distant from the other two, but the values of Q of Dixon 0.717 and of G of Grubbs 1.110 are lower than tabulated values for $P = 0.05$, that is, $Q_{tab} = 1.155$ and $G_{tab} = 1.15$ (although the G_{exp} and G_{tab} values are practically the same). The mean ± confidence limits of the values are 15.093 ± 0.217 mL for hydrochloric acid (first endpoint) and 14.417 ± 0.038 mL for acetic acid (difference), which leads to molarity values of the solutions of hydrochloric and acetic acids of 0.01497 ± 0.00022 M and 0.01430 ± 0.00004 M. If the most distant values were

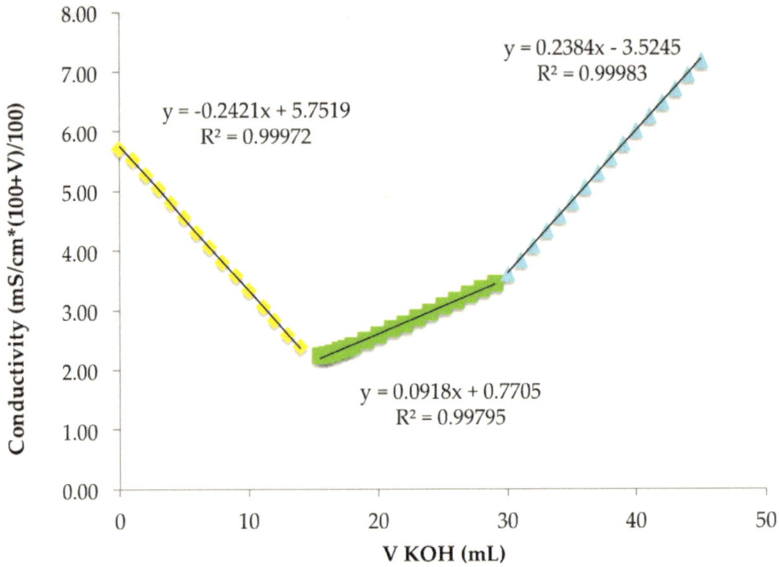

Figure 9. Conductometric titration of a mixture of hydrochloric and acetic acids with potassium hydroxide (data are shown in **Table 10**, second assay).

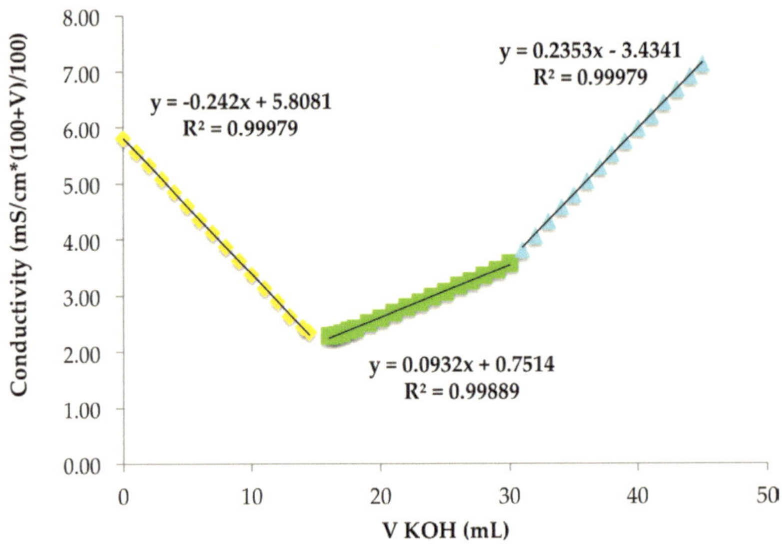

Figure 10. Conductometric titration of a mixture of hydrochloric and acetic acids with potassium hydroxide (data are shown in **Table 9**, third assay).

discarded, the results obtained would be very close to 14.973 ± 0.084 M and 14.421 ± 0.053 M, although the accuracy would improve considerably in the first case.

It is worth noting the fact that when the covariance between the intercept and slope of the straight lines obtained by the least squares method is not taken into account, the propagation

N_1=	13	[ΣW_i]1=	13	N_2=	15	[ΣW_i]2=	15
x_1(mean)=	7.008	y_1(mean)=	4.0562	x_2(mean)=	22.200	y_2(mean)=	2.8029
		[$S(xx)$]1=	182.8092			[$S(xx)$]2=	244.9000
		[$S(xy)$]1=	-44.7709			[$S(xy)$]2=	22.9477
		[$S(yy)$]1=	10.9651			[$S(yy)$]2=	2.1514
b_1=	−0.24491		Δb=	0.33861		b_2=	0.09370
a_1=	5.77243		Δa=	−5.04972		a_2=	0.72272
[$R2$]1=	0.99996		$x(I)$=	14.9132		[$R2$]2=	0.99948
$V[y/x]1$=	4.175E−05					$V[y/x]2$=	8.57549E−05
$V(b_1)$=	2.284E−07		V[pooled]=	6.5584E−05		$V(b_2)$=	3.502E−07
$V(a_1)$=	1.443E−05					$V(a_2)$=	1.783E−04
$cov(a_1, b_1)$=	−1.600E−06					$cov(a_2, b_2)$=	−7.774E−06
$V[\Delta a]$=	1.927E−04	$V[\Delta a]$=	1.590E−04	$V[x(I)]$=	0.0004	FIELLER	$ax2 + bx + c = 0$
$V[\Delta b]$=	5.785E−07	$V[\Delta b]$=	6.266E−07	$s[x(I)]$=	0.0200	a=	1.147E−01
$cov(\Delta a, \Delta b)$=	−9.374E−06	$cov(\Delta a, \Delta b)$=	−8.459E−06	$t(0, 05, 24)$=	2.064	b=	−3.420E+00
		Pooled variances		$t\ s[x(I)]$=	0.0414	c=	2.550E+01
				$Vol\ (u)$=	14.955	$Vol\ (u)$=	14.955
				$Vol\ (l)$=	14.872	$Vol\ (l)$=	14.872
N_3=	15	[ΣW_i]1=	15				
x_3(media)=	37.000	y_3(media)=	5.3001		Δb=	0.145555	
		[$S(xx)$]1=	280.0000		Δa=	−4.275169	
		[$S(xy)$]1=	66.9920		$x(II)$=	29.3716	
		[$S(yy)$]1=	16.0306		Δx=	14.4584	
	b_3=	0.2393			V[pooled]=	0.000132481	
	a_3=	−3.5525					
	[$R2$]3=	0.9999					
	$V[y/x]3$=	1.792E−04		$V[x(II)]$=	0.0034	$t(0, 05, 26)$=	2.056
	$V(b_3)$=	6.400E−07		$s[x(II)]$=	0.0587	$t\ s[x(II)]$=	0.1207
	$V(a_3)$=	8.881E−04				$Vol\ (u)$=	29.492
	$cov(a_3, b_3)$=	−2.368E−05				$Vol\ (l)$=	29.251
				$V[x(I)]$=	4.017E−04		
				$V[x(II)]$=	3.447E−03		
				$Cov(\Delta a_1, \Delta a_2)$=	7.014E−03		
				$Cov(\Delta a_1, \Delta b_2)$=	−9.265E−03		
				$Cov(\Delta a_2, \Delta b_1)$=	−4.704E−03		
				$Cov(\Delta b_1, \Delta b_2)$=	6.224E−03		
$V[\Delta a]$=	1.583E−07	$V[\Delta a]$=	9.320E−04	$V[\Delta x]$=	3.116E−03	$t(0, 05, 37)$=	2.026

N_1=	13	[ΣW_i]1=	13	N_2=	15	[ΣW_i]2=	15
$V[\Delta b]$=	9.902E−07	$V[\Delta b]$=	1.014E−06	$s[\Delta x]$=	5.583E−02	$t\ s[\Delta x]$=	0.113
$cov(\Delta a, \Delta b)$=	−2.893E−05	$cov(\Delta a, \Delta b)$=	−2.952E−05			$Vol\ (u)$=	14.571
		Pooled variances				$Vol\ (l)$=	14.345

Table 13. Evaluation of endpoints in the titration of a mixture of HCl and CH_3COOH with KOH 0.0992 M (data **Table 11**).

of the error leads to values of much larger confidence limits, 0.429 in the example of Massart (1997) versus 0.104, or 0.648 by Liteanu and Rica [58] versus only 0.113, in this book chapter, for the same data. As in many monographs, the covariance in the propagation of errors is not taken into account, and this is perhaps the reason why the estimates of the uncertainties of the intersection abscissa in the analytical literature do not abound.

9. Final comments

The advance of instrumental methods of endpoint detection increases the importance and the worth of titrimetric analysis. Physicochemical methods are intensively developed nowadays. However, titration continues to maintain its importance for chemical analysis. Plotting two straight line graphs from experimental data i.e., the conductivity versus volume added and determining the corresponding intersection point of the two branches allow locating the endpoint in a conductometric titration. The estimation of uncertainty of endpoint from linear segmented titration curves may be easily carried out by first-order propagation of variance, that is, by applying random error propagation law. The weighted linear regression procedure as being applied to the two branches of the conductometric titration curves leads to results similar to those obtained by the unweighted (single) linear regression procedure. The weighting factors are very similar to each other.

The covariance of measurements can be as important as the variance and both contribute significantly to the total analytical error. In particular, the strong correlation existing between the estimated slope and intercept of a straight line obtained by the least squares method must not be ignored. The inclusion of the covariance term on this respect is of vital importance, being usually a subtractive character lowering, in this case, the confidence limits of the abscissa of the intersection point. Perhaps this omission, which leads to too greater uncertainties, may be the cause for a small number of times that uncertainty is reported in this context.

The algebra associated with the Fieller's theorem is simple, and no problem is observed with its derivation in this particular case of intersecting straight lines. However, the statistical uncertainty of endpoint differences is a complex problem. Attempt to derive the confidence limits by applying Fieller's theorem fails in this case, being necessary to resort to the first-order propagation of variance (random error propagation law).

Nevertheless, the algebra associated in this case is simple but cumbersome, as some terms in covariance need to be derived. As a matter of fact, greater accuracy and firmer statistical justification make first-order propagation of variance (random error propagation law) and Fieller's theorem methods preferable to methods based on intersecting confidence bands.

Author details

Julia Martin, Gabriel Delgado Martin and Agustin G. Asuero*

*Address all correspondence to: asuero@us.es

Department of Analytical Chemistry, Faculty of Pharmacy, The University of Seville, Seville, Spain

References

[1] Beck CM. Classical analysis. A look at the past, present and future. Analytical Chemistry. 1994;**66**(4):224A–239A

[2] Granholm K, Sokalski T, Lewenstam A, Ivaska A. Ion-selective electrodes in potentiometric titrations: A new method for processing and evaluating titration data. Analytica Chimica Acta. 2015;**888**:36–43

[3] Thordarson P. Determining association constants from titration experiments in supramolecular chemistry. Chemical Society Reviews. 2011;**40**(3):1305–1323

[4] Winkler-Oswatitsch R and Eigen M. The art of titration. From classical end points to modern differential and dynamic analysis. Angewandte Chemie International Edition English. 1979;**18**(1):20–49

[5] Asuero AG and Michalowski T. Comprehensive formulation of titration curves for complex acid-base systems and its analytical Implications. Critical Reviews in Analytical Chemistry. 2011;**41**(2):151–187

[6] Asuero AG. Buffer capacity of a polyprotic acid: First derivative of the buffer capacity and pKa values of single and overlapping equilibria. Critical Reviews in Analytical Chemistry. 2007;**37**:269–301

[7] Beck CM. Toward a revival of classical analysis. Metrología. 1997;**34**(1):19–30

[8] Felber H, Rezzonico S, Máriássy M. Titrimetry at a metrological level. Metrología. 2003;**40**(5): 249–254

[9] King B. Review of the potential of titrimetry as a primary method. Metrología. 1997;**34**(1):77–82

[10] Ortiz-Fernandez MC, Herrero-Gutierrez A. Regression by least median squares, a methodological contribution to titration. Chemometrics and Intelligent Laboratory Systems. 1995;**27**:241–243

[11] Kupka K, Meloun M. Data analysis in the chemical laboratory II. The end-point estimation in instrumental titrations by nonlinear regression. Analytica Chimica Acta. 2001;**429**:171–183

[12] Jones RH, Molitoris BA. A statistical method for determining the breakpoint of two lines. Analytical Biochemistry. 1984;**41**(1):287–290

[13] Sprent P. Some hypothesis concerning two phase regression analysis. Biometrics. 1961;**17**(4):634–645

[14] Shaban SA. Change point problem and two-phase regression: An annotated bibliography. International Statistical Review. 1980;**48**(1):83–93

[15] Shanubhogue A, Rajarshi MB, Gore AP, Sitaramam V. Statistical testing of equality of two break points in experimental data. Journal of Biochemical and Biophysical Methods. 1992;**25**(2-3):95–112; Erratum 1994;**28**(1): 83

[16] Csörgo M, Horváth L. Nonparametric methods for changepoint problems. In Krishanaiah PR, Rao CR, editors, Handbook of Statistics, Vol. 7, Amsterdam: Elsevier, 1988. pp. 403–425

[17] Krishanaiah PR, Miao BQ. Review about estimation of change points. In Krishanaiah PR, Rao CR, editors. Handbook of Statistics. Vol. 7, Amsterdam: Elsevier, 1988. pp. 375–402

[18] Rukhin AL. Estimation and testing for the common intersection point. Chemometrics and Intelligent Laboratory Systems. 2008;**90**(2):116–122

[19] Yanagimoto T, Yamamoto E. Estimation of safe doses: Critical review of the Hockey Stick regression method. Environmental Health Perspectives. 1979;**32**:193–199

[20] Kita F, Adam W, Jordam P, Nau WM, Wirz J. 1,3-Cyclopentanedyl diradicals: Substituent and temperature dependence of triplet-singlet intersystem crossing. Journal of the American Chemical Society. 1999;**121**(40):9265–9275

[21] Vieth E. Fitting piecewise linear regression functions to biological responses. Journal of Applied Physiology. 1989;**67**(1):390–396

[22] Piegorsch WW. Confidence intervals on the joint point in segmented regression. Biometrical Unit Technical Reports BU-785-M. 1982

[23] Bacon DW, Watts DG, Estimating the transition between two intersecting straight lines. Biometrika. 1972;**58**(3):525–534

[24] Christensen R. Plane answers to complex questions. The Theory of Linear Models. 4th ed., New York: Springer, 2011

[25] Lee ML, Poon WY, Kingdon HS. A two-phase linear regression model for biologic half-life data. Journal of Laboratory and Clinical Medicine. 1990;**115**(6):745–748

[26] Seber GAF. Linear regression analysis. 7.6 Two-phase linear regression. New York: Wiley, 1977. pp. 205–209

[27] Seber GAF, Lee AJ. Linear regression analysis. 6.5 Two-phase linear regression. 2nd ed., New York: Wiley, 2003. pp. 159–161

[28] Cook DA, Charnock JS. Computer-assisted analysis of functions which may be represented by two intersecting straight lines. Journal of Pharmacological Methods. 1979;**2**(1):13–19

[29] Hubálovsky S. Processing of experimental data as educational method of development of algorithmic thinking. Procedia - Social and Behavioral Sciences. 2015;**191**:1876–1880

[30] Hubálovsky S, Sedivú J. Algorithm for estimates determination of parameters of regression straight polyline with objective elimination of outliers. 2nd International Conference on Information Technology (ICIT) 2010 Gdansk, 28–30 June 2010. pp. 263–266

[31] Ouvrard C, Berthelot M, Lamer T, Exner O. A program for linear regression with a common point of intersection. Journal of Chemical Information and Modeling. 2001;**4**(5):1141–1144

[32] Jehlicka V, Mach V. Determination of intersection of regression straight lines with elimination of outliers. Chemical Papers. 1999;**53**(5):279–282

[33] Jehlicka V, Mach V. Determination of estimates of parameters of calibration straight line with objective elimination of remote measurements. Collect. Collection of Czechoslovak Chemical Communications. 1995;**60**(12):2064–2073

[34] Kastenbaum MA. A confidence interval on the abscissa of the point of intersection of two fitted straight lines. Biometrics. 1959;**15**(3):323–324

[35] Filliben JJ, McKinney JE. Confidence limits for the abscissa of intersection of two linear regression. Journal of Research of NIST. 1972;**76**B(3-4):179–192

[36] Meier PC, Zünd RE. Statistical methods in analytical chemistry. 2nd ed., New York: Wiley, 2000. pp. 127–128, 374

[37] Miller JN, Miller JC. Statistics and Chemometrics in Analytical Chemistry. Harlow, England: Pearson, 2010. pp. 140–141

[38] Fieller EC. The biological standardization of insulin. Journal of the Royal Statistical Society. 1940–1941;**7**(1):1–64

[39] Fieller EC. A fundamental formula in the statistics of biological assay, and some applications. Quarterly Journal of Pharmacy and Pharmacology. 1944;**17**:117–123

[40] Fieller EC. Some problems in interval estimation. Journal of the Royal Statistical Society: Series B. 1954;**16**(2):175–185

[41] Koschat MA. A characterization of the Fieller solution. Annals of Statistics. 1987;**15**:462–468

[42] Wallace DL. The Behrens-Fischer and Fieller-Creasy problems. In Berger J, Fienberg S, Gani J, Krickeberg K, Olkin I. R.A. Fischer, an appreciation, editors, Berlin: Springer-Verlag, 1980

[43] Wilkinson GN. On resolving the controversy in statistical inference. Journal of the Royal Statistical Society: Series B. 1977;**39**(2):119–171

[44] Cook DA, Charnot JS. Computer assisted analysis of functions which may be represented by two intersecting straight lines. Journal of Pharmacological and Toxicological Methods. 1979;**2**(1):13–19

[45] Han MH. Non-linear Arrhenius plots in temperature-dependent kinetic studies of enzyme reactions. I. Single transition processes. Journal of Theoretical Biology. 1972;**35**(3):543–568

[46] Puterman ML, Hrboticky N, Innist SM. Nonlinear estimation of parameters in biphasic Arrhenius plots. Analytical Biochemistry. 1988;**170**(2):409–420

[47] Baxter DC. Evaluation of the simplified generalised standard additions method for calibration in the direct analysis of solid samples by graphite furnace atomic spectrometric techniques. Journal of Analytical Atomic Spectrometry. 1989;**4**(5):415–421

[48] Bonate PL. Approximate confidence intervals in calibration using the bootstrap. Analytical Chemistry. 1993;**65**(10):1367–1372

[49] Mandel J, Linning FJ. Study of accuracy in chemical analysis using linear calibration curves. Analytical Chemistry. 1957;**29**(5):743–749

[50] Schwartz LM. Nonlinear calibration curves. Analytical Chemistry. 1977;**49**(13):2062–2066

[51] Schwartz LM. Statistical uncertainties of analyses by calibration of counting measurements. Analytical Chemistry. 1978;**50**(7):980–984

[52] Schwartz LM. Calibration curves with nonuniform variance. Analytical Chemistry. 1979;**51**(6):723–727

[53] Almanda Lopez E, Bosque-Sendra JM, Cuadros Rodriguez L, García Campaña AM, Aaron JJ. Applying non-parametric statistical methods to the classical measurement of inclusion complex binding constants. Analytical and Bioanalytical Chemistry. 2003;**375**(3):414–423

[54] Asuero AG, Recamales MA. A bilogarithmic method for the spectrophotometric evaluation of acidity constants of two-step overlapping equilibria. Analytical Letters 1993;**26** (1):163–181

[55] Heilbronner E. Position and confidence limits of an extremum. The determination of the absorption maximum in wide bands. Journal of Chemical Education. 1979;**56**(4):240–243

[56] Franke JP, de Zeeuw RA, Hakkert R. Evaluation and optimization of the standard addition method for absorption spectrometry and anodic stripping voltammetry. Analytical Chemistry. 1978;**50**(9):1374–1380

[57] Liteanu C, Rica I, Liteanu V. On the confidence interval of the equivalence point in linear titrations. Talanta. 1978;**25**(10):593–596

[58] Liteanu C, Rica I. Statistical Theory and Methodology of Trace Analysis. Chichester: Ellis Horwood Ltd, 1980. pp. 166–172

[59] Asuero AG, Sayago A, González AG. The correlation coefficient: an overview. Critical Reviews in Analytical Chemistry. 2006;**36**(1):41–59

[60] Asuero AG, Gonzalez G, de Pablos F, Ariza JL. Determination of the optimum working range in spectrophotometric procedures. Talanta. 1988;**35**(7):531–537

[61] Ceasescu D, Eliu-Ceasescu VZ. Die Sicherheit der Ergebnisse im Falle von linearen physikalisch-chemischen Titrationskurven. Das Vertrauensintervall des Äquivalenzpunktes. Zeitschrift für Analytische Chemie. 1976;**282**(1):45–46

[62] Ceasescu D, Eliu-Ceasescu V. Equivalence point and confidence interval of linear titration curves in outlook of normal bidimensional distribution of branches. Revue Roumaine de Chimie. 1977;**22**(4):563–567

[63] Draper NR, Smith H. Applied Regression Analysis. New York: Wiley, 1998

[64] McCormick D, Roach A. Measurement, Statistics and Computation, ACOL. Chichester: Wiley, 1987. pp. 312–315

[65] Schwartz LM, Gelb RI. Statistical uncertainties of end points at intersecting straight lines. Analytical Chemistry. 1984;**56**(8):1487–1492

[66] Sayago A, Asuero, A.G. Fitting straight lines with replicated observations by linear regression: Part II. Testing for homogeneity of variances. Critical Reviews in Analytical Chemistry. 2004;**34**(3–4):133–146

[67] Sharaf MA, Illman DL, Kowalski BR. Chemometrics. New York: Wiley, 1986

[68] Brownlee KA. Statistical Theory and Methodology in Science and Engineering. 2nd ed., Malabar, FL: R.E. Krieger, 1984

[69] Carter KN, Scott DM, Salmon JK, Zarcone GS. Confidence limits for the abscissa of intersection of two least squares lines such as linear segmented titration curves. Analytical Chemistry. 1991;**63**(13):1270–1278

[70] Eliu-Ceasescu V, Ceasescu D. Statistische untersuchung des Äquivalenzpunktes vonlinearen Titrationskurven mit Hilfe numerischer Simulierung. Fres. Zeitschrift für Analytische Chemie. 1978;**291**(1):42–46

[71] Jandera P, Kolda S, Kotrly S. End-point evaluation in instrumental titrimetry-II Confidence intevals in extrapolation of linear titration curves. Talanta. 1970;**17**(6):443–454

[72] Liteanu C, Cormos D. Beiträge zum Problem der äquivalenzpunktbestimmung. VII Über die Präzision der äquivalenzpunktbestimmung bei der physikalisch-chemischen linearen titration mit Hilfe der analytischen Methode und der kleinsten Quadrate. Revue Roumaine de Chimie. 1965;**10**:361–376

[73] Sharaf MA, Illman DL, Kowalski BR. Chemometrics. New York: Wiley, 1986

[74] Gelhaus SL, Lacourse WR. Measurement of Electrolytic Conductance. In Analytical Instrumentation Handbook. New York: Marcel Dekker, 2005. pp. 561–580

[75] Gzybkovski W. Conductometric and potentiometric titrations. Politechnika Gdańska, Gdansk, 2002. http://fizyczna.chem.pg.edu.pl/documents/175260/14212622/chf_epm_lab_1.pdf [accessed on August 5, 2017]

[76] Cáñez-Carrrasco MG, García-Alegría AM, Bernal-Mercado AT, Federico-Pérez RA, Wicochea-Rodríguez JD. Conductimetría y titulaciones, ¿cuando, por qué y para qué?. Educación química. 2011;**22**(2):166–169

[77] Donkersloot MCA. Teaching conductometry. Another perspective. Journal of Chemical Education. 1991;**68**(2):136–137

[78] Garcia J, Schultz LD. Determination of sulphate by conductometric titration: an undergraduate laboratory experiment. Journal of Chemical Education. 2016;**93**(5):910–914

[79] Smith KC, Garza A. Using conductivity measurements to determine the identities and concentrations of unknown acids: an inquiry laboratory experiment. Journal of Chemical Education. 2015;**92**(8):1373–1377

[80] Compton OC, Egan M, Kanakaraj R, Higgins TB, Nguyen ST. Conductivity through polymer electrolytes and its implications in lithium-ion batteries: real-world application of periodic trends. Journal of Chemical Education. 2012;**89**(11):1442–1446

[81] Farris S, Mora L, Capretti G, Piergiovanni L. Charge density quantification of polyelectrolyte polysaccharides by conductimetric titration: an analytical chemistry experiment. Journal of Chemical Education. 2012;**89**(1):121–124

[82] Nyasulu F, Moehring M, Arthasery P, Barlag R. K_a and K_b from pH and conductivity measurements: a general chemistry laboratory exercise. Journal of Chemical Education. 2011;**88**(5):640–642

[83] Smith KC, Edionwe E, Michel B. Conductimetric titrations: a predict - observe -explain activity for general chemistry. Journal of Chemical Education. 2010;**87**(11):1217–1221

[84] Holdsworth DK. Conductivity titrations - a microcomputer approach. Journal of Chemical Education. 1986;**63**(1):73–74

[85] Rosenthal LC, Nathan LC. A conductimetric-potentiometric titration for an advanced laboratory. Journal of Chemical Education. 1981;**58**(8):656–658

[86] Rodríguez-Laguna N, Rojas-Hernández A, Ramírez-Silva MT, Hernández-García L, Romero-Romo M. An exact method to determine the conductivity of aqueous solutions in acid-base titrations. Journal of Chemistry. 2015, Article ID 540368, pp. 13

[87] Selitrenikov AV, Zevatskii Yu E. Study of acid-base properties of weak electrolytes by conductometric titration. Russian Journal of General Chemistry. 2015;**85**(1):7–13

[88] Apelblat A, Bester-Rogac M, Barthel J, Neueder R. An analysis of electrical conductances of aqueous solutions of polybasic organic acids. Benzenehexacarboxylic (mellitic) acid and its neutral and acidic salts. Journal of Physical Chemistry B. 2006;**110**(17):8893–8906

[89] Massart D, Vandeginste B, Buydens L, De Jong S, Lewi PJ, Smeyers-Verbeke J. Handbook of Chemometrics and Qualimetrics: Part A, no. 20A in Data Handling in Science and Technology. Amsterdam: Elsevier Science, 1997. p. 305

Section 2

Isothermal Titration Calorimetry

Chapter 4

Determination of Thermodynamic Partial Properties in Multicomponent Systems by Titration Techniques

Mónica Corea, Jean-Pierre E. Grolier and
José Manuel del Río

Additional information is available at the end of the chapter

http://dx.doi.org/10.5772/intechopen.69706

Abstract

Thermodynamic experimental techniques using titration are usually employed to study the interaction between solutes in a diluted solution. This chapter deals with the underlying thermodynamic framework when titration technique is applied with densimetry, sound speed measurement and isothermal titration calorimetry. In the case of partial volumes and partial adiabatic compressibilities, a physical interpretation is proposed based upon atomic, free volume and hydration contributions.

Keywords: thermodynamics, molar partial volumes, molar partial adiabatic compressibilities, molar partial enthalpies, densimetry, sound velocity, isothermal titration calorimetry

1. Introduction

The purposes of this chapter are twofold. First, the thermodynamics fundaments are studied in detail to determine experimentally, calculate and interpret thermodynamic partial molar properties using different titration techniques. Second, the postgraduate students are provided with the necessary thermodynamic background to extract behavioural trends from experimental techniques including densimetry, sound speed measurement and isothermal titration calorimetry.

The first concept introduced in this chapter is "thermodynamic description". It is defined as a set of variables employed to define thermodynamically the studied system. For example, a description by components of a multicomponent system is:

$$J = J(n_1, n_2, n_3) \tag{1}$$

where J is an extensive thermodynamic property; n_1, n_2 and n_3 are the number of moles of components 1, 2 and 3. Other type of thermodynamic description is in terms of the concept of "fraction of a system". A fraction of a system is a thermodynamic entity, with internal composition, which groups several components. For example, the above-mentioned system can be considered as being composed of the component 1, and a fraction F grouping components 2 and 3. In this way, J can be written as:

$$J = J(n_1, n_F, x_{f3}) \tag{2}$$

where n_F is the total number of moles of the fraction F and x_{f3} is a variable related to the composition of the fraction. Depending on the system, one can choose the more adequate description. For example, in a liquid mixture, a description by components (Eq. (1)) can be suitable. Other systems as those shown in **Figure 1** could be better described in terms of fractions.

Figure 1A shows a system composed of the solvent (component 1), solute A (component 2) and solute B (component 3). This system will be described in this chapter using a description by fractions representing a "complex solute" composed of solutes A and B (see **Figure 1B**). This description is appropriate to use in conditions of infinite dilution and dilute solutions. Other example (see **Figure 1C** and **D**) is a functionalized latex particle. A latex is a system composed of polymeric particles dispersed in a solvent. In a functionalized latex, particles are composed of non-polar groups and functional groups (usually polar groups). In this case, a description by components expressed in Eq. (1) and visualized in **Figure 1C** is very difficult to use and it is more convenient to consider a fraction (polymeric particle) composed of non-polar groups (component 2) and polar groups (component 3). **Figure 1D** shows a sketch of this description.

When different descriptions are considered for a system, we have to reconsider the relation between the description and the thermodynamic object studied. In principle, one might think that all descriptions are equivalent. But this is not true because not all descriptions can retain all

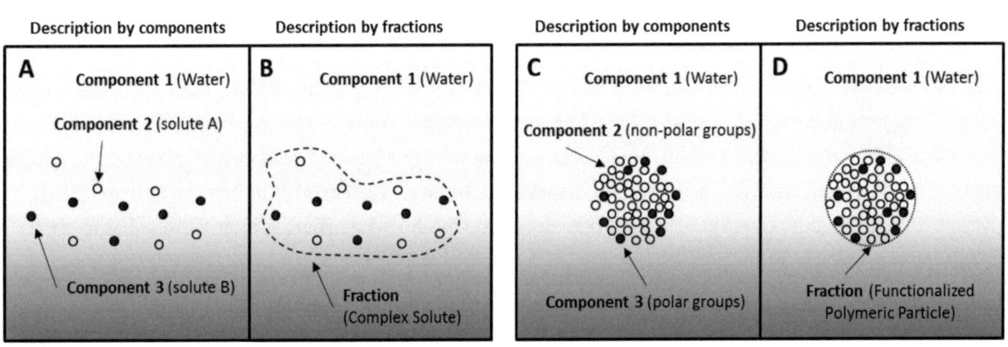

Figure 1. Examples of different descriptions in two systems. (A) and (B) are several solutes in a solvent. (C) and (D) are a functionalized latex with polar groups.

features of a thermodynamic system. For example, it is not possible to speak about thermodynamic partial properties at infinite dilution in multicomponent systems. This fact should not be surprising because in differential geometry [1], there is the same problem associated with the relation between a parametrization and a geometric object. Let's consider, for example, the sphere of radius equal to one, and a parametrization is:

$$\mathbf{X}_1(x,y) = \left(x, y, \sqrt{1-(x^2+y^2)}\right) \tag{3}$$

The problem with this parametrization is that it only covers the top half of the sphere. In addition, it is not differentiable in the points of the sphere's equator. Other possibility is:

$$\mathbf{X}_2(x,y) = \left(x, y, -\sqrt{1-(x^2+y^2)}\right) \tag{4}$$

But, it only covers the lower half of the sphere and neither is differentiable in points of the sphere's equator. Even if we consider a combination of $\mathbf{X}_1$ and $\mathbf{X}_2$, we have the problem of the lack of differentiability in the points of the sphere's equator. Another possible parametrization is:

$$\mathbf{X}_3(\theta,\varphi) = (\sin\theta\cos\varphi,\ \sin\theta\sin\varphi,\ \cos\theta) \tag{5}$$

where θ is the colatitude (the complement of the latitude) and φ the longitude. $\mathbf{X}_3$ covers the whole surface of the sphere and it is also differentiable in all points. For this reason, it contains more information about the sphere (geometric object) than $\mathbf{X}_1$ and $\mathbf{X}_2$. Backing to thermodynamics, in the same case than for $\mathbf{X}_3$, the partial molar properties at infinite dilution cannot be obtained and manipulated using the description by components, and it is necessary to use the description by fractions.

The other concept also introduced in this chapter is the "interaction between components of a system". The first principle of thermodynamics establishes the way, in which systems interact between them and/or with surroundings. In this case, we are interested in the interaction inside the systems and this cannot be interpreted macroscopically using the first principle of thermodynamics. With the concept of interaction between components, we can define mathematically a dilute solution and characterize its thermodynamic behaviour in terms of molar partial properties. In addition to this, we will consider the partial molar properties at infinite dilution. These properties are essential in studies of polymeric particles because they contain the information about the interactions inside the particles. These interactions determine the architecture and final application of the particle.

2. Mathematical fundaments

In this section, some mathematical tools are presented such as changes of variable, changes of size, the Euler theorem and limits in multivariable functions. Variable changes will allow us to relate partial properties of different descriptions. Changes of size are the processes underlying the extensivity and non-extensivity of thermodynamic properties, which will be mathematically implemented by the concept of homogeneity. The Euler's theorem will be treated in the more general form, and in its demonstration we will avoid some aspects, which remain unclear in the versions of the textbooks of Callen [2] and Klotz and Rosenberg [3].

2.1. Changes of variable

Let f be the function defined as:

$$f = f(x_1, x_2, x_3) \tag{6}$$

The gradient of f with respect to the variables x_1, x_2 and x_3 is the vector:

$$\nabla f(x_1, x_2, x_3) = \begin{bmatrix} \left(\frac{\partial f}{\partial x_1}\right)_{x_2, x_3} \\ \left(\frac{\partial f}{\partial x_2}\right)_{x_1, x_3} \\ \left(\frac{\partial f}{\partial x_3}\right)_{x_1, x_2} \end{bmatrix} \tag{7}$$

If we consider the change of variable:

$$\begin{cases} x_1 = x_1(y_1, y_2, y_3) \\ x_2 = x_2(y_1, y_2, y_3) \\ x_3 = x_3(y_1, y_2, y_3) \end{cases} \tag{8}$$

the function f will take the form:

$$f = f(y_1, y_2, y_3) \tag{9}$$

where its gradient will be:

$$\nabla f(y_1, y_2, y_3) = \begin{bmatrix} \left(\frac{\partial f}{\partial y_1}\right)_{y_2, y_3} \\ \left(\frac{\partial f}{\partial y_2}\right)_{y_1, y_3} \\ \left(\frac{\partial f}{\partial y_3}\right)_{y_1, y_2} \end{bmatrix} \tag{10}$$

Our interest is to relate the partial derivatives with respect to the variables x_1, x_2 and x_3 given in Eq. (7) with the partial properties with respect to y_1, y_2 and y_3 given in Eq. (10). From Eq. (8), the total differential of x_1 is:

$$dx_1 = \left(\frac{\partial x_1}{\partial y_1}\right)_{y_2, y_3} dy_1 + \left(\frac{\partial x_1}{\partial y_2}\right)_{y_1, y_3} dy_2 + \left(\frac{\partial x_1}{\partial y_3}\right)_{y_1, y_2} dy_3 \tag{11}$$

Using dx_1 given by Eq. (11) and similarly with equations for dx_2 and dx_3, we can write:

$$\begin{bmatrix} dx_1 \\ dx_2 \\ dx_3 \end{bmatrix} = \mathbf{T}\begin{pmatrix} x_1\ x_2\ x_3 \\ y_1\ y_2\ y_3 \end{pmatrix}\begin{bmatrix} dy_1 \\ dy_2 \\ dy_3 \end{bmatrix} \tag{12}$$

where the matrix **T** is:

$$\mathbf{T}\begin{pmatrix} x_1\ x_2\ x_3 \\ y_1\ y_2\ y_3 \end{pmatrix} = \begin{bmatrix} \left(\frac{\partial x_1}{\partial y_1}\right)_{y_2,y_3} & \left(\frac{\partial x_1}{\partial y_2}\right)_{y_1,y_3} & \left(\frac{\partial x_1}{\partial y_3}\right)_{y_1,y_2} \\ \left(\frac{\partial x_2}{\partial y_1}\right)_{y_2,y_3} & \left(\frac{\partial x_2}{\partial y_2}\right)_{y_1,y_3} & \left(\frac{\partial x_2}{\partial y_3}\right)_{y_1,y_2} \\ \left(\frac{\partial x_3}{\partial y_1}\right)_{y_2,y_3} & \left(\frac{\partial x_3}{\partial y_2}\right)_{y_1,y_3} & \left(\frac{\partial x_3}{\partial y_3}\right)_{y_1,y_2} \end{bmatrix} \tag{13}$$

From (6) and using (7), the total differential of f can be expressed as:

$$df = \left(\frac{\partial f}{\partial x_1}\right)_{x_2,x_3} dx_1 + \left(\frac{\partial f}{\partial x_2}\right)_{x_1,x_3} dx_2 + \left(\frac{\partial f}{\partial x_3}\right)_{x_1,x_2} dx_3 = [\nabla f(x_1,x_2,x_3)]^{\mathrm{T}} \begin{bmatrix} dx_1 \\ dx_2 \\ dx_3 \end{bmatrix} \tag{14}$$

where the symbol "T" indicates "transpose". From Eq. (9) using Eq. (10), the differential of f can be written as:

$$df = \left(\frac{\partial f}{\partial y_1}\right)_{y_2,y_3} dy_1 + \left(\frac{\partial f}{\partial y_2}\right)_{y_1,y_3} dy_2 + \left(\frac{\partial f}{\partial y_3}\right)_{y_1,y_2} dy_3 = [\nabla f(y_1,y_2,y_3)]^{\mathrm{T}} \begin{bmatrix} dy_1 \\ dy_2 \\ dy_3 \end{bmatrix} \tag{15}$$

Equaling (15) to (14) and using (12):

$$[\nabla f(y_1,y_2,y_3)]^{\mathrm{T}} = [\nabla f(x_1,x_2,x_3)]^{\mathrm{T}}\mathbf{T}\begin{pmatrix} x_1\ x_2\ x_3 \\ y_1\ y_2\ y_3 \end{pmatrix} \tag{16}$$

Remembering that **x** being a vector and **A** a matrix, then $(x^{\mathrm{T}} A)^{\mathrm{T}} = A^{\mathrm{T}} x$, and taking the transpose in both sides of (16):

$$\nabla f(y_1, y_2, y_3) = \left[\mathbf{T}\begin{pmatrix} x_1\ x_2\ x_3 \\ y_1\ y_2\ y_3 \end{pmatrix}\right]^{\mathrm{T}} \nabla f(x_1, x_2, x_3) \tag{17}$$

Eq. (17) relates the vector gradient with respect to the variables (x_1, x_2, x_3) to the vector gradient with respect to the variables (y_1, y_2, y_3), and it will allow us to express the partial properties in two different descriptions.

2.2. Changes of size

In this paragraph, the process of size change in thermodynamic systems is analyzed. The behaviour of systems in a size change has consequences on the behaviour or nature of the thermodynamic properties as well as on the form of the thermodynamic equations of the system. **Figure 2** shows a visualization of this process in both directions: increasing and reduction.

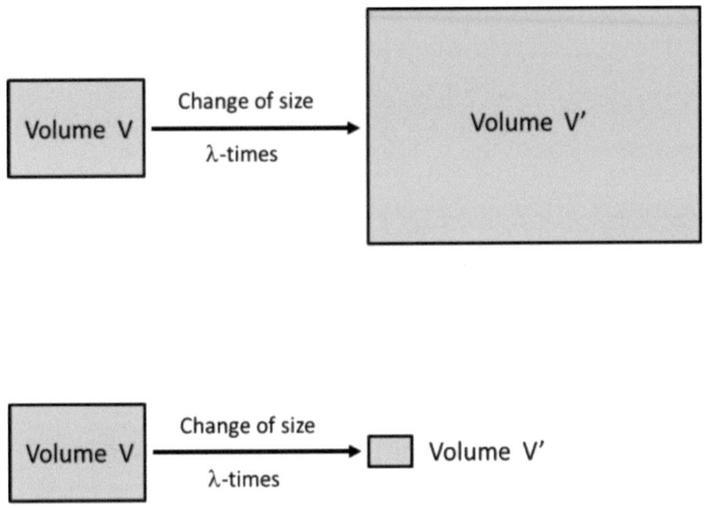

Figure 2. Sketch of the change of size (increasing and reduction) of a system with volume *V*.

From **Figure 2**, it is clear that being V the volume, N the number of moles and U the internal energy, the configuration of this system is under increasing size λ times:

$$\begin{aligned} & V \xrightarrow{\lambda-times} V' = \lambda V \\ & N \xrightarrow{\lambda-times} N' = \lambda N \\ & U \xrightarrow{\lambda-times} U' = \lambda U \end{aligned} \tag{18}$$

Thermodynamic properties, which transform accordingly to (18), depend on the size of the system and are named extensive variables. Not all thermodynamic variables transform according to Eq. (18). An example is the molar fraction of the component 2 (x_2) in a two-component system. We can see this formally in the following way. For a two-component system:

$$\begin{aligned} & N_1 \xrightarrow{\lambda-times} N'_1 = \lambda N_1 \\ & N_2 \xrightarrow{\lambda-times} N'_2 = \lambda N_2 \end{aligned} \tag{19}$$

and x_2 transforms as:

$$x_2 \xrightarrow{\lambda-times} x'_2 = \frac{N'_2}{N'_1 + N'_2} = \frac{\lambda N_2}{\lambda N_1 + \lambda N_2} = \frac{N_2}{N_1 + N_2} = x_2 \tag{20}$$

That is, the molar fraction of the component 2 is independent of the system size. Properties, which remain constant upon size change, are named intensive properties. Other thermodynamic properties with such characteristics are temperature, pressure, pH and concentration c_2 ($c_2 = N_2/V$). It is also interesting to look at the behaviour of functions, which depend on thermodynamic variables (intensive and/or extensive), in a size change. Let, for example, the function f be given by $f = f(T, P, N_1, N_2, \ldots)$. For particular values of the variables T_0, P_0, N_{01}, N_{02}, …, the function f takes the value f_0, and in a change of size:

$$\begin{aligned} &T_0 \xrightarrow{\lambda-times} T'_0 = T_0 \\ &P_0 \xrightarrow{\lambda-times} P'_0 = P_0 \\ &N_{01} \xrightarrow{\lambda-times} N'_{01} = \lambda N_{01} \\ &N_{02} \xrightarrow{\lambda-times} N'_{02} = \lambda N_{02} \\ &\dots \\ &f_0 \xrightarrow{\lambda-times} f'_0 \end{aligned} \tag{21}$$

If $f'_0 = lf_0$, the function will behave as an extensive property. This concept is mathematically implemented as:

$$f(T, P, \lambda N_1, \lambda N_2, \dots) = \lambda f(T, P, N_1, N_2, \dots) \tag{22}$$

In this case, f is a homogeneous function of one degree. If $f'_0 = f_0$, f will behave as an intensive property. Mathematically, f is expressed as a homogeneous function of zero degree as:

$$f(T, P, \lambda N_1, \lambda N_2, \dots) = f(T, P, N_1, N_2, \dots) \tag{23}$$

2.3. Euler's theorem

Let $f = f(x_1, x_2, \dots; y_1, y_2, \dots)$ be a function, which is a homogeneous function of one degree with respect to the variables $y_1, y_2, \dots$:

$$f(x_1, x_2, \dots; \lambda y_1, \lambda y_2, \dots) = \lambda f(x_1, x_2, \dots; y_1, y_2, \dots) \tag{24}$$

Then,

$$f = \left(\frac{\partial f}{\partial y_1}\right)_{x_1, x_2, \dots; y_2, y_3, \dots} y_1 + \left(\frac{\partial f}{\partial y_2}\right)_{x_1, x_2, \dots; y_1, y_3, \dots} y_2 + \dots. \tag{25}$$

The demonstration is as follows. The differential with respect to λ in the left side of (24) is:

$$\begin{aligned} \frac{df(x_1, x_2, \dots; \lambda y_1, \lambda y_2, \dots)}{d\lambda} = &\left(\frac{\partial f(x_1, x_2, \dots; \lambda y_1, \lambda y_2, \dots)}{\partial x_1}\right)_{x_2, x_3, \dots; \lambda y_1, \lambda y_2, \dots} \\ &\frac{dx_1}{d\lambda} + \left(\frac{\partial f(x_1, x_2, \dots; \lambda y_1, \lambda y_2, \dots)}{\partial x_2}\right)_{x_1, x_3, \dots; \lambda y_1, \lambda y_2, \dots} \\ &\frac{dx_2}{d\lambda} + \left(\frac{\partial f(x_1, x_2, \dots; \lambda y_1, \lambda y_2, \dots)}{\partial (\lambda y_1)}\right)_{x_1, x_2, \dots; \lambda y_2, \lambda y_3, \dots} \\ &\frac{d(\lambda y_1)}{d\lambda} + \left(\frac{\partial f(x_1, x_2, \dots; \lambda y_1, \lambda y_2, \dots)}{\partial (\lambda y_2)}\right)_{x_1, x_2, \dots; \lambda y_1, \lambda y_3, \dots} \frac{d(\lambda y_2)}{d\lambda} + \dots \end{aligned} \tag{26}$$

For the sets of variables $x_1, x_2, \dots$ and $y_1, y_2, \dots$, we obtain respectively that:

$$\frac{dx_1}{d\lambda} = \frac{dx_2}{d\lambda} = \ldots = 0 \tag{27}$$

$$\frac{d(\lambda y_1)}{d\lambda} = y_1, \qquad \frac{d(\lambda y_2)}{d\lambda} = y_2, \ \ldots \tag{28}$$

The following step in this demonstration is different from the step proposed in other textbooks [2, 3]. The partial derivative of f with respect to (λy_1) can be expressed as:

$$\left(\frac{\partial f(x_1, x_2, \ldots; \lambda y_1, \lambda y_2, \ldots)}{\partial(\lambda y_1)}\right)_{x_1, x_2, \ldots; \lambda y_2, \lambda y_3, \ldots} = \lim_{\Delta\to 0} \frac{f(x_1, x_2, \ldots; \lambda y_1 + \Delta, \lambda y_2, \ldots) - f(x_1, x_2, \ldots; \lambda y_1, \lambda y_2, \ldots)}{\Delta} \tag{29}$$

Considering that f is a homogeneous function of one degree with respect to the variables y_1, y_2, … and making $\Delta' = \Delta/\lambda$ in (29),

$$\left(\frac{\partial f(x_1, x_2, \ldots; \lambda y_1, \lambda y_2, \ldots)}{\partial(\lambda y_1)}\right)_{x_1, x_2, \ldots; \lambda y_2, \lambda y_3, \ldots} = \lim_{\Delta'\to 0} \frac{f(x_1, x_2, \ldots; y_1 + \Delta', y_2, \ldots) - f(x_1, x_2, \ldots; y_1, y_2, \ldots)}{\Delta'} = \left(\frac{\partial f(x_1, x_2, \ldots; y_1, y_2, \ldots)}{\partial y_1}\right)_{x_1, x_2, \ldots; y_2, y_3, \ldots} \tag{30}$$

The differential of f with respect to λ in the right side of (24) is:

$$\frac{d[\lambda f(x_1, x_2, \ldots; y_1, y_2, \ldots)]}{d\lambda} = f(x_1, x_2, \ldots; y_1, y_2, \ldots) \tag{31}$$

Eq. (25) is obtained by substituting Eqs. (27), (28), (30), (31) in Eq. (26). In addition, it is interesting to see that, defining f_1 as $f_1 = (\partial f/\partial x_1)$ and using (30), f_1 is a homogeneous function of zero degree with respect to the variables y_1, y_2,…:

$$f_1(x_1, x_2, \ldots; \lambda y_1, \lambda y_2, \ldots) = f_1(x_1, x_2, \ldots; y_1, y_2, \ldots) \tag{32}$$

3. Thermodynamic descriptions

3.1. Description by components

Let it be a three-component system (e.g., as those of **Figure 1A** and **C**). Being J an extensive property, a description by components is:

$$J = J(n_1, n_2, n_3) \tag{33}$$

where n_1, n_2 and n_3 are the number of moles of components 1, 2 and 3. The partial property of 1 is defined as:

http://dx.doi.org/10.5772/intechopen.69706

$$j_{1;2,3}(n_1, n_2, n_3) = \left(\frac{\partial J(n_1, n_2, n_3)}{\partial n_1}\right)_{n_2, n_3} \tag{34}$$

From the above section, we know that $j_{1;2,3}$ is homogeneous function of zero degree with respect to n_1, n_2 and n_3. With this and considering $\lambda = 1/(n_1+n_2+n_3)$,

$$\begin{aligned} j_{1;2,3}(n_1, n_2, n_3) &= j_{1;2,3}\left(\frac{n_1}{n_1+n_2+n_3}, \frac{n_2}{n_1+n_2+n_3}, \frac{n_3}{n_1+n_2+n_3}\right) = \\ &= j_{1;2,3}(x_1, x_2, x_3) = j_{1;2,3}(x_2, x_3) \end{aligned} \tag{35}$$

where we have considered that x_1 is a function of x_2 and x_3 because $x_1 = 1-x_2-x_3$. From (35), we see that the partial molar properties depend only on the composition of the system. Alternatively to (35), we could use other scales of composition/concentration to express $j_{1;2,3}$.

The equation of Gibbs is obtained by differentiating J in (33) and using Eq. (34) and similar definitions for components 2 and 3:

$$dJ = j_{1;2,3}dn_1 + j_{2;1,3}dn_2 + j_{3;1,2}dn_3 \tag{36}$$

The Euler equation is obtained by considering that J is a homogeneous function with respect to n_1, n_2 and n_3 and applying the Euler's theorem:

$$J = n_1 j_{1;2,3} + n_2 j_{2;1,3} + n_3 j_{3;1,2} \tag{37}$$

The Gibbs-Duhem equation is obtained by differentiating in Eq. (37), equalling to Eq. (36) and cancelling common terms:

$$0 = n_1 dj_{1;2,3} + n_2 dj_{2;1,3} + n_3 dj_{3;1,2} \tag{38}$$

If we consider that partial molar properties are function of n_1, n_2 and n_3, Eq. (38) would be the Gibbs-Duhem equation in the representation of variables n_1, n_2 and n_3. The representation in the variables x_2 and x_3 is as follows. Dividing (38) by the total number of moles,

$$0 = x_1 dj_{1;2,3} + x_2 dj_{2;1,3} + x_3 dj_{3;1,2} \tag{39}$$

Calculating the differentials by considering that partial molar properties depend on x_2 and x_3, and bearing in mind that x_2 and x_3 are independent variables, (38) can be written in an alternative way as:

$$\begin{cases} x_1\left(\frac{\partial j_{1;2,3}}{\partial x_2}\right)_{x_3} + x_2\left(\frac{\partial j_{2;1,3}}{\partial x_2}\right)_{x_3} + x_3\left(\frac{\partial j_{3;1,2}}{\partial x_2}\right)_{x_3} = 0 \\ x_1\left(\frac{\partial j_{1;2,3}}{\partial x_3}\right)_{x_2} + x_2\left(\frac{\partial j_{2;1,3}}{\partial x_3}\right)_{x_2} + x_3\left(\frac{\partial j_{3;1,2}}{\partial x_3}\right)_{x_2} = 0 \end{cases} \tag{40}$$

3.2. Description by fractions

In a description by fractions, we consider the three-component system as composed of a component 1 and a group (or fraction) composed of components 2 and 3. **Figure 1B** shows the example when two solutes are grouped in a "complex solute", and **Figure 1D** shows the example in which a polymeric particle composed of polar and non-polar groups is considered as a fraction of the system. In this case, the extensive property J is expressed as:

$$J = J(n_1, n_F, x_{f3}) \tag{41}$$

where

$$n_F = n_2 + n_3 \tag{42}$$

$$x_{f3} = \frac{n_3}{n_2 + n_3} \tag{43}$$

The variable n_F is the total number of moles of the fraction F, and x_{f3} is a variable related to its internal composition. The partial molar properties of J in this description are:

$$j_{1;F}(n_1, n_F, x_{f3}) = \left(\frac{\partial J(n_1, n_F, x_{f3})}{\partial n_1}\right)_{n_F, x_{f3}} \tag{44}$$

$$j_{F;1}(n_1, n_F, x_{f3}) = \left(\frac{\partial J(n_1, n_F, x_{f3})}{\partial n_F}\right)_{n_1, x_{f3}} \tag{45}$$

Because J is a homogeneous function of n_1 and n_F, the partial properties $j_{1;F}$ and $j_{F;1}$ will be homogeneous functions of zero degree with respect to the variables n_1 and n_F. In this way and similarly to Eq. (35):

$$j_{1;F}(n_1, n_F, x_{f3}) = j_{1;F}(x_F, x_{f3}) \tag{46}$$

where $x_F = n_F/(n_1+n_F)$. Now, we will see the relation between both descriptions. From (42) and (43), the change of variable of Eq. (8) is in this case:

$$\begin{cases} n_1(n_1, n_F, x_{f3}) = n_1 \\ n_2(n_1, n_F, x_{f3}) = (1 - x_{f3})n_F \\ n_3(n_1, n_F, x_{f3}) = x_{f3}n_F \end{cases} \tag{47}$$

Substituting (47) in (13) and the result in (17), one obtains that:

$$j_{1;F} = j_{1;2,3} \tag{48}$$

$$j_{F;1} = x_{f2}j_{2;1,3} + x_{f3}j_{3;1,2} \tag{49}$$

$$\left(\frac{\partial J}{\partial x_{f3}}\right)_{n_1, n_F} = n_F(j_{3;1,2} - j_{2;1,3}) \tag{50}$$

The equations of Gibbs, Euler and Gibbs-Duhem in this description are as follows. The Gibbs equation is obtained by differentiating in (41) and considering the definitions given in (44) and (45):

$$dJ = j_{1;F}dn_1 + j_{F;1}dn_F + \left(\frac{\partial J}{\partial x_{f3}}\right)_{n_1, n_F} dx_{f3} \tag{51}$$

The Euler equation is obtained by remembering that J is a homogeneous function of degree one of n_1 and n_F and using the Euler's theorem:

$$J = n_1 j_{1;F} + n_F j_{F;1} \tag{52}$$

The Gibbs-Duhem equation in the representation of variables x_F and x_{f3} is obtained by differentiating in (52), equalling to (51) and cancelling common terms, and dividing by the total number of moles:

$$x_1 dj_{1;F} + x_F dj_{F;1} = x_F(j_{3;1,2} - j_{2;1,3})dx_{f3} \tag{53}$$

Considering that $j_{1;F}$ and $j_{F;1}$ are functions of the independent variables x_F and x_{f3}, then (53) will take the form:

$$\begin{cases} x_1\left(\frac{\partial j_{1;F}}{\partial x_F}\right)_{x_{f3}} + x_F\left(\frac{\partial j_{F;1}}{\partial x_F}\right)_{x_{f3}} = 0 \\ x_1\left(\frac{\partial j_{1;F}}{\partial x_{f3}}\right)_{x_F} + x_F\left(\frac{\partial j_{F;1}}{\partial x_{f3}}\right)_{x_F} = x_F(j_{3;1,2} - j_{2;1,3}) \end{cases} \tag{54}$$

Calculating the partial derivative of $j_{F;1}$ with respect to x_{f3} in Eq. (49) and substituting in Eq. (54), we obtain:

$$\begin{cases} x_1\left(\frac{\partial j_{1;F}}{\partial x_F}\right)_{x_{f3}} + x_F\left(\frac{\partial j_{F;1}}{\partial x_F}\right)_{x_{f3}} = 0 \\ x_1\left(\frac{\partial j_{1;F}}{\partial x_{f3}}\right)_{x_F} + x_F\left[x_{f2}\left(\frac{\partial j_{2;1,3}}{\partial x_{f3}}\right)_{x_F} + x_{f3}\left(\frac{\partial j_{3;1,2}}{\partial x_{f3}}\right)_{x_F}\right]_{x_F} = 0 \end{cases} \tag{55}$$

It is interesting to observe that considering constant composition ($dx_{f3} = 0$) in Eqs. (51)–(53), then the system behaves as a two-component system. This fact cannot be obtained using the description by components.

4. Partial properties in diluted solutions of multicomponent systems

We consider intuitively a diluted solution when the properties of the solution are similar to those of its solvent in pure state. In this section, we will study the thermodynamic behaviour of the partial molar properties in this region of concentrations.

4.1. Thermodynamic concept of interaction between components

In this paragraph, we will define the concept of non-interaction and prove that when applying it to a system, the system behaves as an ideal mixing. From a thermodynamic point of view, the components of a system are not interacting if both following points hold simultaneously.

1. The state of each component in the system, expressed in terms of its partial molar properties, does not vary by changes of composition of the other components. It means each component does not detect the presence of the other components.

2. The formation of the system from its pure components is carried out with any cost of energy, neither for the system nor for the surroundings.

Mathematically, the first point can be written as:

$$\left(\frac{\partial j_{1;2,3}(x_2,x_3)}{\partial x_2}\right)_{x_3}=\left(\frac{\partial j_{1;2,3}(x_2,x_3)}{\partial x_3}\right)_{x_2}=0\Leftrightarrow j_{1;2,3}=j_{1;2,3}(x_1) \tag{56}$$

Substituting (56) in (40) and considering also that:

$$\begin{cases}\left(\frac{\partial j_{1;2,3}}{\partial x_2}\right)_{x_3}=\left(\frac{\partial j_{1;2,3}}{\partial x_1}\right)_{x_3}\left(\frac{\partial x_1}{\partial x_2}\right)_{x_3}=-\left(\frac{\partial j_{1;2,3}}{\partial x_1}\right)_{x_3}\\ \left(\frac{\partial j_{1;2,3}}{\partial x_3}\right)_{x_2}=\left(\frac{\partial j_{1;2,3}}{\partial x_1}\right)_{x_2}\left(\frac{\partial x_1}{\partial x_3}\right)_{x_2}=-\left(\frac{\partial j_{1;2,3}}{\partial x_1}\right)_{x_2}\end{cases} \tag{57}$$

it is obtained that:

$$\begin{cases}-x_1\left(\frac{\partial j_{1;2,3}}{\partial x_1}\right)_{x_3}+x_2\left(\frac{\partial j_{2;1,3}}{\partial x_2}\right)_{x_3}=0\\ -x_1\left(\frac{\partial j_{1;2,3}}{\partial x_1}\right)_{x_2}+x_3\left(\frac{\partial j_{2;1,3}}{\partial x_3}\right)_{x_2}=0\end{cases} \tag{58}$$

Because $j_{1;2,3}$ depends only on x_1:

$$\left(\frac{\partial j_{1;2,3}}{\partial x_1}\right)_{x_2}=\left(\frac{\partial j_{1;2,3}}{\partial x_1}\right)_{x_3} \tag{59}$$

and then (57) yields:

$$x_1\left(\frac{\partial j_{1;2,3}(x_1)}{\partial x_1}\right)_{x_3}=x_2\left(\frac{\partial j_{2;1,3}(x_2)}{\partial x_2}\right)_{x_3}=x_3\left(\frac{\partial j_{3;1,2}(x_3)}{\partial x_3}\right)_{x_3} \tag{60}$$

Because the first term depends only on x_1 and the second and third terms depend only on x_2 and x_3, respectively, from (60), we have that:

$$x_1\frac{dj_{1;2,3}}{dx_1}=k_J(T,P) \tag{61}$$

where k_J is a function, which only depends on temperature T and pressure P. Similar equations to (61) are obtained for $j_{2;1,3}$ and $j_{3;1,2}$. Integrating in (61) between $x'_1=1$ and x_1,

$$j_{1;2,3}(x_1)=j_1+k_J(T,P)\ln(x_1) \tag{62}$$

http://dx.doi.org/10.5772/intechopen.69706

For components 2 and 3, similar equations to (62) are obtained. Now, we will apply the second point of the above definition of non-interaction. The zero cost of energy for the system and surroundings is equivalent to:

$$Q_{mix} = W_{mix} = 0 \Rightarrow \Delta U_{mix} = 0 \Rightarrow \begin{cases} u_{1;2,3} = u_1 \\ h_{1;2,3} = h_1 \\ v_{1;2,3} = v_1 \end{cases} \tag{63}$$

Considering $u_{1;\,2,3}$, $h_{1;\,2,3}$ and $v_{1;\,2,3}$ as (62) and bearing in mind (63):

$$k_U(T,P) = k_H(T,P) = k_V(T,P) = 0 \tag{64}$$

In addition to this $g_{1;2,3}$ (free energy of Gibbs),

$$v_{1;2,3} = \left(\frac{\partial g_{1;2,3}}{\partial P}\right)_{T,x_2,x_3} \Rightarrow k_V(T,P) = \left(\frac{\partial k_G(T,P)}{\partial P}\right)_T \tag{65}$$

$$\frac{h_{1;2,3}}{T^2} = -\left(\frac{\partial}{\partial T}\left(\frac{g_{1;2,3}}{T}\right)\right)_{P,x_2,x_3} \Rightarrow \frac{k_H(T,P)}{T^2} = -\left(\frac{\partial}{\partial T}\left(\frac{k_G(T,P)}{T}\right)\right)_T \tag{66}$$

Combining Eqs. (64)–(66), we have that:

$$k_G(T,P) = kT \tag{67}$$

where k is a constant. For the entropy, one gets:

$$-s_{1;2,3} = \left(\frac{\partial g_{1;2,3}}{\partial T}\right)_{P,x_2,x_3} \Rightarrow -k_S(T,P) = \left(\frac{\partial k_G(T,P)}{\partial T}\right)_P \tag{68}$$

With this,

$$g_{1;2,3} = g_1 + kT\ln(x_1) \tag{69}$$

$$s_{1;2,3} = s_1 - k\ln(x_1) \tag{70}$$

and we have demonstrated that a system holding the non-interaction definition proposed is an ideal mixing.

4.2. Diluted solutions

In this section, we will define the thermodynamic concept of diluted solutions and study the behaviour of the partial molar properties of these solutions. Commonly and intuitively, we consider a solution as diluted when its properties are similar to those of the pure solvent. We can implement mathematically this concept in the following way. When we remove all solutes from a solution, we have that:

$$\lim_{x_2+x_3\to 0} j(x_2,x_3) = j_1 \tag{71}$$

where j is the molar property of the extensive thermodynamic property J. In addition, the partial derivatives must vanish:

$$\lim_{x_2+x_3\to 0}\left(\frac{\partial j_{1;2,3}(x_2,x_3)}{\partial x_2}\right)_{x_3}=\lim_{x_2+x_3\to 0}\left(\frac{\partial j_{1;2,3}(x_2,x_3)}{\partial x_3}\right)_{x_2}=0 \tag{72}$$

Otherwise, we would have memory effects and we can see this with an example. If we purify water, the pure substance obtained does not depend on the initial diluted solution employed. Actually, pure water is commonly used as a standard because it does not depend on the part of world, in which it is obtained. The Taylor's expansion of $j_{1;2,3}$ is:

$$j_{1;2,3}(x_2,x_3)=j_1(0,0)+\left[\nabla j_{1;2,3}(0,0)\right]^{\mathrm{T}}\begin{bmatrix}x_2\\x_3\end{bmatrix}+\frac{1}{2}(x_2,x_3)\mathbf{H}j_{1;2,3}(0,0)\begin{bmatrix}x_2\\x_3\end{bmatrix}+\dots \tag{73}$$

where $\nabla j_{1;2,3}(0,0)$ and $\mathbf{H}j_{1;2,3}(0,0)$ are, respectively, the vector gradient and the Hessian of $j_{1;2,3}$ matrix at (0,0). Considering (71) and (72) in (73) and that all partial derivatives mush vanish at (0,0), we have that for diluted solutions:

$$j_{1;2,3}(x_2,x_3)\approx j_1+\dots \tag{74}$$

From Eq. (74), we have for diluted solutions:

$$\left(\frac{\partial j_{1;2,3}}{\partial x_2}\right)_{x_3}\approx\left(\frac{\partial j_{1;2,3}}{\partial x_3}\right)_{x_2}\approx 0 \tag{75}$$

The behaviour of molar partial properties of solutes is as follows. Considering a "complex solute" S composed of 2 and 3 (as in **Figure 1B**),

$$j_{1;S}(x_S,x_{s3})\approx j_1 \tag{76}$$

and substituting Eq. (74) in the first equation of (55),

$$\left(\frac{\partial j_{S;1}}{\partial x_S}\right)_{x_{s3}}\approx 0\Leftrightarrow j_{S;1}(x_S,x_{s3})\approx j_{S;1}(x_{s3}) \tag{77}$$

Inserting Eq. (76) in the second equation of (55), it is obtained that:

$$x_{s2}\left(\frac{\partial j_{2;1,3}}{\partial x_{s3}}\right)_{x_S}+x_{s3}\left(\frac{\partial j_{3;1,2}}{\partial x_{s3}}\right)_{x_S}\approx 0 \tag{78}$$

Until now, we have seen the effect of the dilution in the capacity of detecting the presence of other components in a diluted solution. In order to gain an insight into the interactions, we have to study the process of mixing in diluted solutions. From (71), we can write:

$$\begin{cases}\lim_{x_2+x_3\to 0} h=h_1\Rightarrow\lim_{x_2+x_3\to 0} q_{mix}=\lim_{x_2+x_3\to 0}\Delta_{mix}h=0\\ \lim_{x_2+x_3\to 0} v=v_1\Rightarrow\lim_{x_2+x_3\to 0} w_{mix}=\lim_{x_2+x_3\to 0}-P\Delta_{mix}v=0\end{cases} \tag{79}$$

It indicates that in the limit of infinite dilution, components do not interact because the process of mixture does not have any energy cost. This result implicates that in diluted solutions, according to the asymptotic approach given by Eq. (74), the interaction between solvent and solutes is weak and it can be neglected.

4.3. Partial molar properties of interaction in diluted solutions

The molar property j of a diluted solution can be written as:

$$j \approx x_1 j_1 + x_S j_{S;1} \tag{80}$$

where we are considering the interaction between components 2 and 3 since

$$j_{S;1} = x_{s2} j_{2;1,3} + x_{s3} j_{3;1,2} \tag{81}$$

In a diluted solution without interaction between 2 and 3, the property $j^{\varnothing}$ can be written as:

$$j^{\varnothing} = x_1 j_1 + x_S (x_{s2} j_{2;1} + x_{s3} j_{3;1}) \tag{82}$$

In this way, we can calculate the interaction contributions to j as:

$$\Delta j^{\text{int}} = j - j^{\varnothing} \approx x_S \Delta j_{S;1} \tag{83}$$

where

$$\Delta j_{S;1} = x_{s2} \Delta j_{2;1,3} + x_{s3} \Delta j_{3;1,2} \tag{84}$$

is the partial molar property of interaction of the complex solute and

$$\begin{cases} \Delta j_{2;1,3} = j_{2;1,3} - j_{2;1} \\ \Delta j_{3;1,2} = j_{3;1,2} - j_{3;1} \end{cases} \tag{85}$$

are the partial molar properties of interaction of the components 2 and 3, respectively. These properties are not independent as we will see as follows. Combining (78) and (85),

$$x_{s2} \left(\frac{\partial \Delta j_{2;1,3}}{\partial x_{s3}} \right)_{x_S} + x_{s3} \left(\frac{\partial \Delta j_{3;1,2}}{\partial x_{s3}} \right)_{x_S} + x_{s2} \left(\frac{\partial j_{2;1}}{\partial x_{s3}} \right)_{x_S} + x_{s3} \left(\frac{\partial j_{3;1}}{\partial x_{s3}} \right)_{x_S} \approx 0 \tag{86}$$

In Eq. (85), $\Delta j_{2;1,3}$ and $\Delta j_{3;1,2}$ are evaluated when using concentrations x_S and x_{s3}. Accordingly, $j_{2,1}$ is evaluated using the concentration x_2 given by $x_2 = x_S\,(1-x_{s3})$, and then,

$$\left(\frac{\partial j_{2;1}}{\partial x_{s3}} \right)_{x_S} = \frac{d j_{2;1}}{d x_2} \left(\frac{\partial x_2}{\partial x_{s3}} \right)_{x_S} = -x_S \frac{d j_{2;1}}{d x_2} \tag{87}$$

Considering the Gibbs-Duhem equation for a two-component system:

$$x_1 \frac{dj_{1;2}}{dx_2} + x_2 \frac{dj_{2;1}}{dx_2} = 0 \tag{88}$$

in Eq. (87) and bearing in mind that solutions are diluted,

$$x_{s2} \left(\frac{\partial j_{2;1}}{\partial x_{s3}} \right)_{x_S} = x_1 \frac{dj_{1;2}}{dx_2} \approx 0 \tag{89}$$

Similarly for component 3,

$$x_{s3} \left(\frac{\partial j_{3;1}}{\partial x_{s3}} \right)_{x_S} = x_1 \frac{dj_{1;3}}{dx_3} \approx 0 \tag{90}$$

Substituting (89) and (90) in (86), we obtain:

$$x_{s2} \left(\frac{\partial \Delta j_{2;1,3}}{\partial x_{s3}} \right)_{x_S} + x_{s3} \left(\frac{\partial \Delta j_{3;1,2}}{\partial x_{s3}} \right)_{x_S} \approx 0 \tag{91}$$

Eq. (91) indicates that in a diluted solution, the interaction between components 2 and 3 is not vanished. The partial molar property of interaction of the complex solute can be calculated experimentally as:

$$\Delta j_{S;1} \approx j_{S;1} - (x_{s2} j_{2;1} + x_{s3} j_{3;1}) \tag{92}$$

and the partial properties of interaction of components 2 and 3 can be obtained from (92) using the equations:

$$\begin{cases} \Delta j_{2;1,3} \approx \Delta j_{S;1} - x_{s3} \dfrac{d\Delta j_{S;1}}{dx_{s3}} \\ \Delta j_{3;1,2} \approx \Delta j_{S;1} + (1 - x_{s3}) \dfrac{d\Delta j_{S;1}}{dx_{s3}} \end{cases} \tag{93}$$

Eq. (93) is obtained by differentiating in Eq. (92) with respect to x_{s3}, using Eq. (91) and combining the result with Eq. (92). As we will see below, Eq. (93) will allow us to obtain the interaction partial properties of 2 and 3 from experimental data.

4.4. Experimental determination of partial molar properties of interaction in diluted solutions

4.4.1. *Partial specific volumes of interaction and partial specific adiabatic compressibility of interaction*

As an example, we will consider the interaction between functionalized polymeric particles and an electrolyte at 30°C [4]. For that, polymeric particles synthesized of poly(n-butyl acrylate-co-methyl methacrylate) functionalized with different concentrations of acrylic acid were used in this study. The electrolyte was NaOH. Similarly to **Figure 1A**, water (solvent) was considered as component 1, polymeric particles as component 2 and electrolyte as component

3. And similarly to **Figure 1B**, the system was fractionalized in component 1 and a complex solute composed of polymeric particles and electrolyte. The experimental measurements were carried out using a Density and Sound Analyzer DSA 5000 from Anton-Paar connected to a titration cell. It is of full cell type, which is usually employed in isothermal titration calorimetry. Polymeric particles were located in the titration cell, and electrolyte was located in the syringe. Concentrations of polymeric particles (c_2) and electrolyte (c_3) after each titration were calculated as [4, 5]:

$$\begin{cases} c_2^{i+1} = c_2^i \, e^{-\frac{v}{V}} \\ c_3^{i+1} = c_3^s - \left(c_3^s - c_3^i\right) e^{-\frac{v}{V}} \end{cases} \tag{94}$$

where V is the effective volume of the titration cell, v is the titration volume and c_3^s is the stock concentration of electrolyte in the syringe. **Figure 3A** and **B** shows, respectively, data of density (ρ) and sound speed (u) as function of the electrolyte concentration. The specific volume (v) and the specific adiabatic compressibility (ks) were calculated as:

$$v = \frac{1}{\rho} \tag{95}$$

$$k_s = \left(\frac{10}{\rho u}\right)^2 \tag{96}$$

Considering the solution in the cell as diluted, the partial specific volume (and similarly the partial specific adiabatic compressibility) of the complex solute can be calculated as:

$$v_{S;1} = \frac{v - t_1 v_1}{t_S} \tag{97}$$

where t_1 and t_S are the mass fraction of the water and of the complex solute, respectively. **Figure 3C** and **D** shows the partial specific volume and partial specific adiabatic compressibility as function of t_{f3} (mass fraction of the electrolyte in the complex solute). The term of interaction $\Delta v_{S;1}$ is calculated by Eq. (92), where $v_{2;1}$ is obtained by considering that:

$$v_{2;1} = \lim_{t_{s3} \to 0} v_{S;1} \tag{98}$$

in **Figure 3C**. The term $v_{3;1}$ is calculated by extrapolating the linear part of $v_{S;1}$ in **Figure 3C** as:

$$v_{3;1} = \lim_{t_{s3} \to 1} v_{S;1} \tag{99}$$

The partial specific volume of interaction of the polymeric particles ($\Delta v_{2;1,3}$) and the partial specific volume of interaction of the electrolyte ($\Delta v_{3;1,2}$) were obtained using Eq. (93). The numerical method employed to calculate the derivatives is shown elsewhere [4]. **Figure 4** shows the values of $\Delta v_{S;1}$, $\Delta v_{2;1,3}$ and $\Delta v_{3;1,2}$, and **Figure 5** shows the values of $\Delta k_{s\ S;1}$, $\Delta k_{s\ 2;1,3}$ and $\Delta k_{s\ 3;1,2}$ obtained in a similar way than for volumes.

Partial volume of polymeric particles ($v_{2;1}$) can be broken down in the following contributions [6–11]:

$$v_{2;1} = v_{2;1/atom} + v_{2;1/free} + v_{2;1/hyd} \tag{100}$$

which are shown in **Figure 6**. The atomic volume contribution ($v_{2;1/\text{atom}}$) is the sum of all volumes of the atoms, which make up polymeric chains. The free volume contribution ($v_{2;1/\text{free}}$) is consequence of the imperfect packing of the polymeric chains. The atomic volume contribution and free volume contribution are both positive contributions. The hydration contribution ($v_{2;1/\text{hyd}}$) is negative, as a consequence of that the specific volume of water molecules in bulk is larger than the specific volume in the hydration shell. The contributions to the partial specific adiabatic compressibility are the free volume and hydration because the effect of the pressure on the atomic volume is neglected [10, 12–21]:

$$k_{T\,2;1} = -\left(\frac{\partial v_{2;1}}{\partial P}\right)_T = -\left(\frac{\partial v_{2;1/free}}{\partial P}\right)_T - \left(\frac{\partial v_{2;1/hyd}}{\partial P}\right)_T = k_{T\,2;1/free} + k_{T\,2;1/hyd} \tag{101}$$

The contribution $k_{T\ 2;1/\text{free}}$ is positive, and the contribution $k_{T\ 2;1/\text{hyd}}$ is negative [4, 8]. In this chapter, we will take the adiabatic compressibility as an approximation of the isothermal

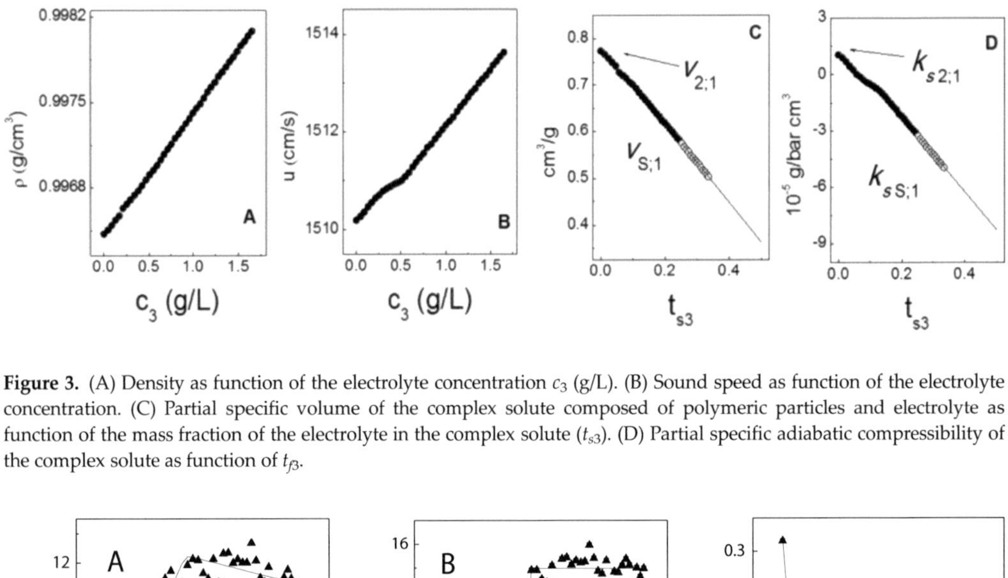

Figure 3. (A) Density as function of the electrolyte concentration c_3 (g/L). (B) Sound speed as function of the electrolyte concentration. (C) Partial specific volume of the complex solute composed of polymeric particles and electrolyte as function of the mass fraction of the electrolyte in the complex solute (t_{s3}). (D) Partial specific adiabatic compressibility of the complex solute as function of t_{f3}.

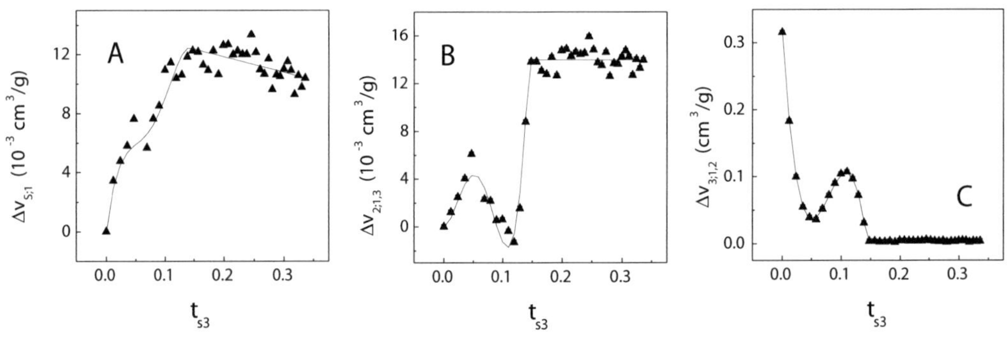

Figure 4. (A) Partial specific volume of interaction of the complex solute (polymeric particles + electrolyte) as function of the mass fraction of the electrolyte in the complex solute (t_{s3}). (B) Partial specific volume of interaction of the polymeric particles as function of t_{s3}. (C) Partial specific volume of interaction of the electrolyte as function of t_{s3}.

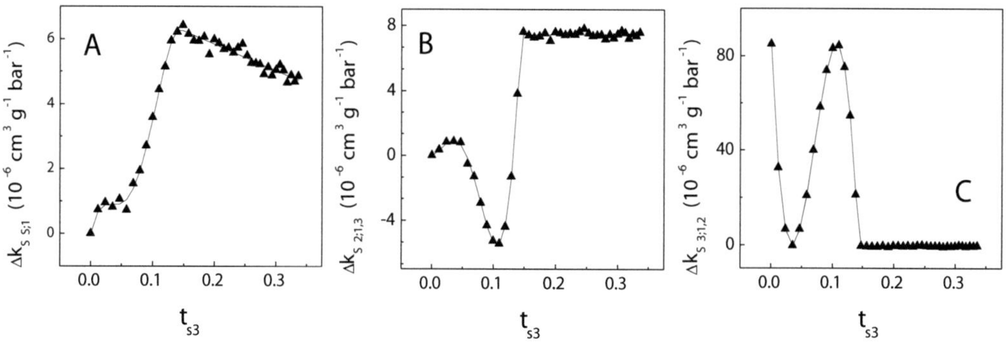

Figure 5. (A) Partial specific adiabatic compressibility of interaction of the complex solute (polymeric particles + electrolyte) as function of the mass fraction of the electrolyte in the complex solute (t_{s3}). (B) Partial specific adiabatic compressibility of interaction of the polymeric particles as function of t_{s3}. (C) Partial specific adiabatic compressibility of interaction of the electrolyte as function of t_{s3}.

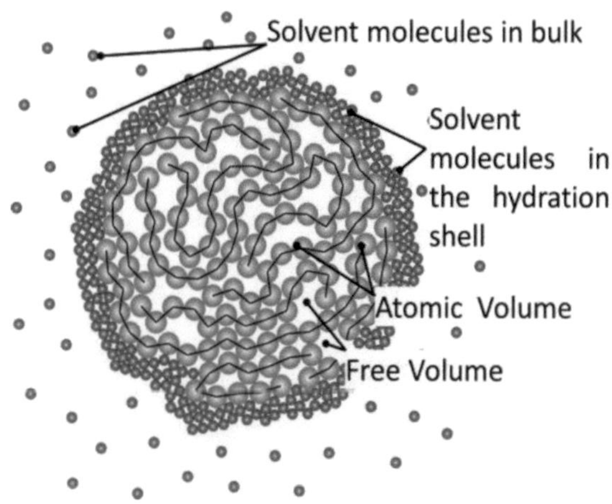

Figure 6. Contributions to the partial volume in a polymeric particle.

compressibility. For the electrolyte, the free volume contribution is null, and then, $v_{3;1}$ and $k_{T\ 3;1}$ will take the following form:

$$v_{3;1} = v_{3;1/atom} + v_{3;1/hyd} \tag{102}$$

$$k_{T3;1} = -\left(\frac{\partial v_{3;1}}{\partial P}\right)_T = -\left(\frac{\partial v_{3;1/hyd}}{\partial P}\right)_T = k_{T3;1/hyd} \tag{103}$$

For the complex solute, we can write a similar breakdown:

$$v_{S;1} = v_{S;1/atom} + v_{S;1/free} + v_{S;1/hyd} \tag{104}$$

Inserting Eqs. (100), (102) and (104) in Eq. (92) and neglecting the variation in the atomic contributions, the following equation for the interaction specific partial volume is obtained:

$$\Delta v_{S,1} = \Delta v_{S,1/free} + \Delta v_{S,1/hyd} \tag{105}$$

where

$$\Delta v_{S;1/free} = v_{S;1/free} - t_{s2} v_{2;1/free} \tag{106}$$

$$\Delta v_{S;1/hyd} = v_{S;1/hyd} - (t_{s2} v_{2;1/hyd} + t_{s3} v_{3;1/hyd}) \tag{107}$$

Substituting (105) in (93), we get

$$\begin{cases} \Delta v_{2;1,3} = \left(\Delta v_{S;1/free} - t_{s3} \dfrac{d\Delta v_{S;1/free}}{dt_{s3}}\right) + \left(\Delta v_{S;1/hyd} - t_{s3} \dfrac{d\Delta v_{S;1/hyd}}{dt_{s3}}\right) \\ \Delta v_{3;1,2} = \left(\Delta v_{S;1/free} + (1 - t_{s3}) \dfrac{d\Delta v_{S;1/free}}{dt_{s3}}\right) + \left(\Delta v_{S;1/hyd} + (1 - t_{s3}) \dfrac{d\Delta v_{S;1/hyd}}{dt_{s3}}\right) \end{cases} \tag{108}$$

Defining now:

$$\begin{aligned} \Delta v_{2;1,3/free} &= \Delta v_{S;1/free} - t_{s3} \frac{d\Delta v_{S;1/free}}{dt_{s3}} \\ \Delta v_{2;1,3/hyd} &= \Delta v_{S;1/hyd} - t_{s3} \frac{d\Delta v_{S;1/hyd}}{dt_{s3}} \\ \Delta v_{3;1,2/free} &= \Delta v_{S;1/free} + (1 - t_{s3}) \frac{d\Delta v_{S;1/free}}{dt_{s3}} \\ \Delta v_{3;1,2/hyd} &= \Delta v_{S;1/hyd} + (1 - t_{s3}) \frac{d\Delta v_{S;1/hyd}}{dt_{s3}} \end{aligned} \tag{109}$$

One arrives at the following result:

$$\begin{aligned} \Delta v_{2;1,3} &= \Delta v_{2;1,3/free} + \Delta v_{S;1/free} \\ \Delta v_{3;1,2} &= \Delta v_{3;1,2/free} + \Delta v_{3;1,2/hyd} \end{aligned} \tag{110}$$

where similar equations are obtained for the interaction partial specific compressibilities.

Considering these contributions, the interpretation of the partial specific volumes of interaction of the particle as function of the electrolyte concentration is as follows. From $t_{f3} = 0$ to around 0.05 (see **Figure 4B**), there is an increment in $\Delta v_{2,1,3}$ which can be interpreted as a gain of free volume by the disentanglement of the polymeric chains. This increment of free volume is accompanied by an increment in the hydrodynamic radius [4]. From around $t_{f3} = 0.05$ to around 0.1, there is a decrement in $\Delta v_{2,1,3}$ due to hydration. In this region of compositions, the separation of polymeric chains allows the entrance of water molecules in the polymeric particle. As a result, the hydrodynamic radius of the particle increases [4]. From around $t_{s3} = 0.1$ to 0.15, $\Delta v_{2,1,3}$ increases sharply. This fact can be interpreted as an increment of the dehydration. Beyond $t_{s3} = 0.15$, $\Delta v_{2,1,3}$ becomes constant, indicating that the interaction between particles and the electrolyte is saturated. Similar regions with similar interpretations are obtained for the partial specific adiabatic compressibility (see **Figure 5B**).

4.4.2. Partial specific enthalpies of interaction

This section deals with the determination of the partial specific enthalpies of interaction of the same system than in the latest example [4]. Partial specific enthalpy of interaction of polymeric particles is:

$$\Delta h_{2;1,3} = h_{2;1,3} - h_{2;1} \tag{111}$$

and the partial specific enthalpy interaction of the electrolyte is:

$$\Delta h_{3;1,2} = h_{3;1,2} - h_{3;1} \tag{112}$$

The partial specific enthalpy of interaction of the electrolyte can be measured by isothermal titration calorimetry using the combination of two experiments [6, 7]. The first experiment is locating the polymeric particles in the cell and the electrolyte in the syringe. The heat per unit of titration volume in an infinitesimal titration is:

$$\frac{dQ^{cd}}{dv} = (\rho^s - c_3^s)h_{1;2,3} + c_3^s h_{3;1,2} - h_v(c_3^s) \tag{113}$$

where ρ^s is the density of the stock solution and $h_v(c_3^s)$ is the enthalpy of the stock solution per unit volume. The second experiment consists of titrating water with the above stock solution, and its heat per unit of titration volume in an infinitesimal titration is:

$$\frac{dQ^c}{dv} = (\rho^s - c_3^s)h_{1;3} + c_3^s h_{1;3} - h_v(c_3^s) \tag{114}$$

The partial specific enthalpy of interaction of the electrolyte is obtained by subtracting (114) from (113), considering Eq. (112), diluted solutions and bearing in mind that $dn_3^2 = c_3^s dv$:

$$\frac{dQ^{cd}}{dn_3^s} - \frac{dQ^c}{dn_3^s} = \Delta h_{3;1,2} \tag{115}$$

Figure 7A shows the experimental values $\Delta h_{3;1,2}$. The partial specific enthalpy of interaction of polymeric particles was calculated by integrating Eq. (91) [7]:

$$\Delta h_{2;1,3}(t_{f3}) = -\int_0^{t_{f3}} \frac{t'_{f3}}{1 - t'_{f3}} \left(\frac{d\Delta h_{3;1,2}}{dt'_{f3}} \right) dt'_{f3} \tag{116}$$

and the values of $\Delta h_{2;1,3}$ are shown in **Figure 7B**. It is very interesting to observe in **Figure 7B** that $\Delta h_{2;1,3}$ is zero from t_{s3} = 0 to around t_{s3} = 0.1. This fact indicates that the changes, which take place in the first two regions in **Figures 4B** and **5B**, are entropic in origin.

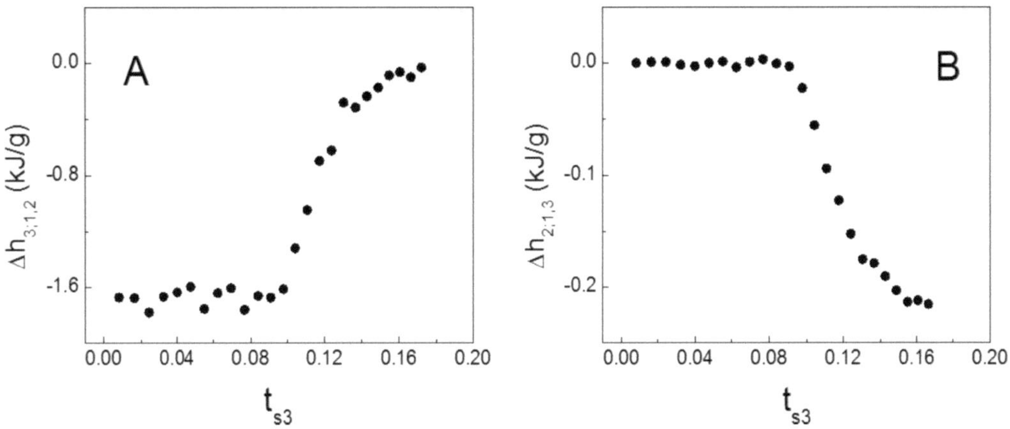

Figure 7. (A) Partial specific enthalpy of interaction of the electrolyte as function of the mass fraction of the electrolyte in the complex solute (t_{s3}). (B) Partial specific enthalpy of interaction of the polymeric particles as function of t_{s3}.

5. Partial molar properties at infinite dilution

First, we will discuss the case of the two-component system and then make the extension to three-component system. In this section, J can be U, H, V or their derivatives $Cv = (\partial H/\partial T)_V$, $C_p = (\partial H/\partial T)_P$ or $E = (\partial V/\partial T)_P$, $K_T = (\partial V/\partial P)_T$ and $K_S = (\partial V/\partial P)_S$.

5.1. Two-component systems

In a two-component system, we only have one way to calculate limits at infinite dilution and it is to take a component as solvent (component 1) and the other as solute (component 2). For a two-component system, j takes the form:

$$j(x_2) = x_1 j_{1;2}(x_2) + x_2 j_{2;1}(x_2) \tag{117}$$

Because

$$\lim_{x_2 \to 0} j(x_2) = j_1 \tag{118}$$

and using Eq. (117), we have:

$$\lim_{x_2 \to 0} j_{1;2}(x_2) = j_1 \tag{119}$$

For the solute, we have:

$$\lim_{x_2 \to 0} j_{2;1}(x_2) = j^o_{2;1} \tag{120}$$

We can obtain experimentally the value of $j^o_{2;1}$ as follows. The Taylor's expansion of $j(x_2)$ around $x_2 = 0$ is:

$$j(x_2) = j(0) + \frac{dj(0)}{dx_2} x_2 + \ldots \tag{121}$$

Differentiating (117) with respect to x_2, considering the Gibbs-Duhem equation for a 2-component system and combining the results with equations (117), (120) and (121):

$$j(x_2) = j_1 + (j^o_{2;1} - j_1)x_2 \tag{122}$$

For this reason, we can obtain experimentally $j^o_{2;1}$ from a linear fit in a plot of $j(x_2)$ as function of x_2.

5.2. Three-component systems

In three-component systems, we have two ways to calculate limits at infinite dilution. The first way is to group two components in a "complex solvent" and to calculate the limit at infinite dilution of the other component in this complex solvent (type I). The other way is considering a component as solvent, to group the other two components in a complex solute, and to calculate the limit at infinite dilution of the complex solute in the solvent (type II).

5.2.1. Limits of type I

In this case, we consider a complex solvent B composed of components 1 and 2 and a solute (component 3). For this system,

$$J = J(n_3, n_B, x_{b2}) \tag{123}$$

where $n_B = n_1+n_2$ and $x_{b2} = n_2/(n_1+n_2)$. With this, j can be written as:

$$j(x_3, x_{b2}) = x_B j_{B;3}(x_3, x_{b2}) + x_3 j_{3;B}(x_3, x_{b2}) \tag{124}$$

where x_3 is the mole fraction of the component 3. At infinite dilution, we have:

$$\lim_{\substack{x_3 \to 0 \\ x_{b2}\ \text{constant}}} j(x_3, x_{b2}) = j_B(x_{b2}) \tag{125}$$

and then combining Eq. (124) with (125), one gets for the solvent:

$$\lim_{\substack{x_3 \to 0 \\ x_{b2}\ \text{constant}}} j_{B;3}(x_3, x_{b2}) = j_B(x_{b2}) \tag{126}$$

For the solute, it is obtained that:

$$\lim_{\substack{x_3 \to 0 \\ x_{b2}\ \text{constant}}} j_{3;B}(x_3, x_{b2}) = j^o_{3;B}(x_{b2}) \equiv j^o_{3;1,2}(x_{b2}) \tag{127}$$

where we have used Eq. (48). Similarly to the case of two-component systems, the amount $j^o_{3;1,2}$ can be obtained experimentally by using the equation:

$$j(x_3, x_{b2}) = j_B(x_{b2}) + \left(j^o_{3;1,2} - j_B(x_{b2})\right)x_3 \tag{128}$$

This equation is obtained by using the first-order Taylor's expansion of $j(x_3,x_{b2})$ around $x_3 = 0$, the partial derivative of $j(x_3, x_{b2})$ with respect to x_3, the Gibbs-Duhem equation of the fractionalized system considering the composition of the fraction as constant and Eqs. (126) and (127).

5.2.2. *Limits of type II*

In this case (see **Figure 1A** and **B**), we will consider the component 1 as solvent and a "complex solute" S composed of 2 and 3 and then:

$$J = J(n_1, n_S, x_{s3}) \tag{129}$$

where $n_S = n_2 + n_3$ and $x_{s3} = n_3/(n_2 + n_3)$. The molar property j is:

$$j(x_S, x_{s3}) = x_1 j_{1;S}(x_S, x_{s3}) + x_S j_{S;1}(x_S, x_{s3}) \tag{130}$$

Similarly to the above cases, at infinite dilution we have for the solvent:

$$\lim_{\substack{x_S \to 0 \\ x_{s3}\ \text{constant}}} j_{1;S}(x_S, x_{s3}) = j_1 \tag{131}$$

Accordingly to case of the two-component system, one gets for the complex solute:

$$\lim_{\substack{x_S \to 0 \\ x_{s3}\ \text{constant}}} j_{S;1}(x_S, x_{s3}) = j^o_{S;1}(x_{s3}) \tag{132}$$

and in a similar way than for the type I limits, $j^o_{S;1}$ can be calculated as

$$j(x_S, x_{s3}) = j_1 + \left(j^o_{S;1}(x_{s3}) - j_1\right) x_S \tag{133}$$

In order to study the contributions of components 2 and 3 to $j^o_{S;1}$, we define the following limits an infinite dilution:

$$\begin{cases} \lim_{\substack{x_S \to 0 \\ x_{s3}\ \text{constant}}} j_{2;1,3}(x_S, x_{s3}) = j^\Delta_{2;1,3}(x_{s3}) \\ \lim_{\substack{x_S \to 0 \\ x_{s3}\ \text{constant}}} j_{3;1,2}(x_S, x_{s3}) = j^\Delta_{3;1,2}(x_{s3}) \end{cases} \tag{134}$$

In this way, taking limits in both sides of Eq. (49), and bearing in mind Eqs. (132) and (134), we have that:

$$j^o_{S;1}(x_{s3}) = x_{s2} j^\Delta_{2;1,3}(x_{s3}) + x_{s3} j^\Delta_{3;1,2}(x_{s3}) \tag{135}$$

Now, we will see some mathematical properties of limits of type II. One of them is for example:

$$\lim_{x_{s3}\to 0} j^\Delta_{2;1,3}(x_{s3}) = j^o_{2;1} \tag{136}$$

This property is demonstrated by using iterated limits:

$$\lim_{x_{s3}\to 0} j^{\Delta}_{2;1,3}(x_{s3}) = \lim_{\substack{x_{s3}\to 0 \\ x_S \text{ constant}}} \left[\lim_{\substack{x_S\to 0 \\ x_{s3} \text{ constant}}} j_{2;1,3}(x_S, x_{s3}) \right]$$

$$= \lim_{\substack{x_S\to 0 \\ x_{s3} \text{ constant}}} \left[\lim_{\substack{x_{s3}\to 0 \\ x_S \text{ constant}}} j_{2;1,3}(x_S, x_{s3}) \right] = \lim_{x_S\to 0} j_{2;1,3}(x_S, 0) = \lim_{x_2\to 0} j_{2;1}(x_2) = j^{o}_{2;1} \tag{137}$$

The other mathematical property is:

$$\lim_{x_{s3}\to 1} j^{\Delta}_{3;1,2}(x_{s3}) = j^{o}_{2;1,3}(0) \tag{138}$$

where its demonstration is as follows:

$$\lim_{x_{s3}\to 1} j^{\Delta}_{3;1,2}(x_{s3}) = \lim_{\substack{x_{s3}\to 1 \\ x_S \text{ constant}}} \left[\lim_{\substack{x_S\to 0 \\ x_{s3} \text{ constant}}} j_{3;1,2}(x_S, x_{s3}) \right]$$
$$= \lim_{\substack{x_S\to 0 \\ x_{s3} \text{ constant}}} \left[\lim_{\substack{x_{s3}\to 1 \\ x_S \text{ constant}}} j_{3;1,2}(x_S, x_{s3}) \right] \tag{139}$$

Now, it is necessary to consider other way to fractionalize the system. For convenience, we will consider a complex solvent B composed of 1 and 3, and a solute 2 where the variable x_B represents the molar fraction of B and $x_{b3} = n_3/(n_1+n_3)$. With this,

$$\begin{cases} \lim_{\substack{x_{s3}\to 1 \\ x_S \text{ constant}}} x_2 = \lim_{\substack{x_{s3}\to 1 \\ x_S \text{ constant}}} x_S(1 - x_{s3}) = 0 \\ \lim_{\substack{x_{s3}\to 1 \\ x_S \text{ constant}}} x_{b3} = \lim_{\substack{x_{s3}\to 1 \\ x_S \text{ constant}}} x_{s3}x_S[1 - x_S(1 - x_{s3})] = x_S \end{cases} \tag{140}$$

and considering Eq. (140), (139) transforms into:

$$\lim_{x_{s3}\to 1} j^{\Delta}_{3;1,2}(x_{s3}) = \lim_{\substack{x_S\to 0 \\ x_{s3} \text{ constant}}} \left[\lim_{\substack{x_2\to 0 \\ x_{b2} \text{ constant}}} j_{3;1,2}(x_2, x_{b2}) \right] =$$
$$= \lim_{\substack{x_S\to 0 \\ x_{s3} \text{ constant}}} j^{o}_{3;1,2}(x_{b2}) = j^{o}_{3;1,2}(0) \tag{141}$$

Other interesting property of the limits of type II is that they are related to each other by the following equation:

$$x_{s2}\frac{dj^{\Delta}_{2;1,3}}{dx_{s3}} + x_{s3}\frac{dj^{\Delta}_{3;1,2}}{dx_{s3}} = 0 \tag{142}$$

The demonstration of this equation is as follows. Both sides of the following equation:

$$x_{s2}\left(\frac{\partial j_{2;1,3}}{\partial x_{s3}}\right)_{x_S}+x_{s3}\left(\frac{\partial j_{3;1,2}}{\partial x_{s3}}\right)_{x_S}=x_S(1-x_S)\left(\frac{\partial (j_{3;1,2}-j_{2;1,3})}{\partial x_S}\right)_{x_{s3}} \tag{143}$$

are calculated in the following way. The left-hand side is obtained by deriving partially Eq. (49) with respect to x_{s3}. The right-hand side of (143) is calculated considering that:

$$\left(\frac{\partial j_{S;1}}{\partial x_{s3}}\right)_{x_S}=\frac{\partial J}{\partial x_{s3}\partial n_S}=\frac{\partial J}{\partial n_S\partial x_{s3}} \tag{144}$$

Using (50) in (144) and cancelling common terms, Eq. (143) is obtained. Taking the limit when x_s approaches to zero when x_{s3} is kept constant in both sides of Eq. (143) and considering that:

$$\lim_{\substack{x_S\to 0\\ x_{s3}\ \text{constant}}}\left(\frac{\partial\left(j_{3;1,2}(x_S,x_{s3})-j_{2;1,3}(x_S,x_{s3})\right)}{\partial x_S}\right)_{x_{s3}}$$
$$=\left(\frac{\partial j_{3;1,2}(0,x_{s3})}{\partial x_S}\right)_{x_{s3}}-\left(\frac{\partial j_{2;1,3}(0,x_{s3})}{\partial x_S}\right)_{x_{s3}}=f(x_{s3}) \tag{145}$$

Eq. (142) is obtained.

From values of $j^{o}_{S;1}$ it is possible to obtain $j^{\Delta}_{2;1,3}$ and $j^{\Delta}_{3;1,2}$ by using the following equations:

$$\begin{cases} j^{\Delta}_{2;1,3}=j^{o}_{S;1}-x_{s3}\dfrac{dj^{o}_{S;1}}{dx_{s3}}\\ j^{\Delta}_{3;1,2}=j^{o}_{S;1}+(1-x_{s3})\dfrac{dj^{o}_{S;1}}{dx_{s3}}\end{cases} \tag{146}$$

Eq. (146) was obtained by differentiating Eq. (135) with respect to x_{s3}, considering Eq. (142) and combining the result with Eq. (135).

5.2.3. *Application of the limits of type II to the study of polymeric particles*

The polymeric particles used were synthesized with a gradient of concentration of functional groups (acrylic acid) inside the particle [9]. In this system, the content of acrylic acid represents the polar groups, while poly(butyl acrylate-co-methylmethacrylate) is the non-polar groups. As seen in **Figure 1C** and **D**, component 1 is water, component 2 is non-polar groups and component 3 is polar groups. The polymeric particle (composed of polar and non-polar groups) is taken as a fraction "P" of the system where the variable $t_{p3}=n_3/(n_2+n_3)$ will be the mass fraction of polar groups in the particle. In this study [9], the same experimental equipment than in Section 4.4.1 was used and measurements of density and sound speed were carried out by titrating water (in the cell) with latex of polymeric particles (in the syringe). **Figure 7A** and **B** shows the density ρ and u as functions of the concentration for several values of t_{p3}. The density and sound speed were transformed into specific volumes and specific adiabatic compressibilities by using Eqs. (95) and (96), and results are shown in **Figure 1C** and **D**.

In this case, Eq. (133) will take the form:

$$j = j_1 + (j^o_{P;1} - j_1)\, t_P \tag{147}$$

and considering that $t_1 = 1 - t_P$, Eq. (147) transforms into:

$$j = j^o_{P;1} + (j_1 - j^o_{P;1})\, t_1 \tag{148}$$

Using Eq. (148) as a fit function in **Figure 8C** and **D**, the partial specific volume at infinite dilution of the particles ($v^o_{P;1}$) and the partial specific adiabatic compressibility at infinite dilution of the particles ($k^o_{S\ P;1}$) were obtained from the independent term of Eq. (148) and the results are shown in **Figure 9A** and **B** as functions of t_{p3}. In this case, Eqs. (100) and (101) take the form:

$$v^o_{P;1} = v^o_{P;1/atom} + v^o_{P;1/free} + v^o_{P;1/hyd} \tag{149}$$

$$k^o_{S\ P;1} = k^o_{S\ P;1/free} + k^o_{S\ P;1/hyd} \tag{150}$$

The partial specific properties of polar ($j^\Delta_{3;1,2}$) and non-polar ($j^\Delta_{2;1,3}$) groups were calculated by using Eq. (146). The derivatives of Eq. (146) were calculated numerically by using schemes of finite differences. **Figure 9C** and **D** shows, as functions of the amount of polar groups (t_{p3}), the values of specific partial volumes of non-polar and polar groups, respectively. **Figure 9D** and **F** shows, respectively, the specific partial adiabatic compressibility of non-polar and polar groups.

With similar arguments than in Section 4.4.1, we can get the following equations for the volumes:

$$v^\Delta_{2;1,3} = v^\Delta_{2;1,3/atom} + v^\Delta_{2;1,3/free} + v^\Delta_{2;1,3/hyd} \tag{151}$$

$$v^\Delta_{3;1,2} = v^\Delta_{3;1,2/atom} + v^\Delta_{3;1,2/free} + v^\Delta_{3;1,2/hyd} \tag{152}$$

and for the adiabatic compressibilities:

$$k^\Delta_{T\ 2;1,3} = k^\Delta_{T\ 2;1,3/free} + k^\Delta_{T\ 2;1,3/hyd} \tag{153}$$

$$k^\Delta_{T\ 3;1,2} = k^\Delta_{T\ 3;1,2/free} + k^\Delta_{T\ 3;1,2/hyd} \tag{154}$$

In addition to this, by combining Eqs. (135), (149), (151) and (152), one gets the following equations:

$$v^o_{P;1/atom} = t_{p2} v^\Delta_{2;1,3/atom} + t_{p3} v^\Delta_{3;1,2/atom} \tag{155}$$

$$v^o_{P;1/free} = t_{p2} v^\Delta_{2;1,3/free} + t_{p3} v^\Delta_{3;1,2/free} \tag{156}$$

$$v^o_{P;1/hyd} = t_{p2} v^\Delta_{2;1,3/hyd} + t_{p3} v^\Delta_{3;1,2/hyd} \tag{157}$$

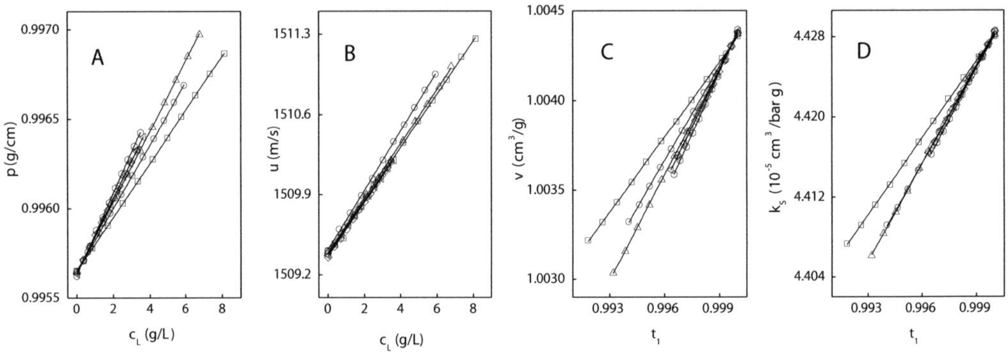

Figure 8. (A) Density of latex as function of polymeric particles concentration. (B) Sound speed as function of polymeric particles concentration. (C) Specific volume of latex as function of mass fraction of solvent (water). (D) Specific adiabatic compressibility as function of the mass fraction of solvent (water). In all figures (□) 0 wt%, (O) 5wt%, (Δ) 10 wt%, (◊) 15 wt %, (◁) 20 wt%, (●) 25wt%.

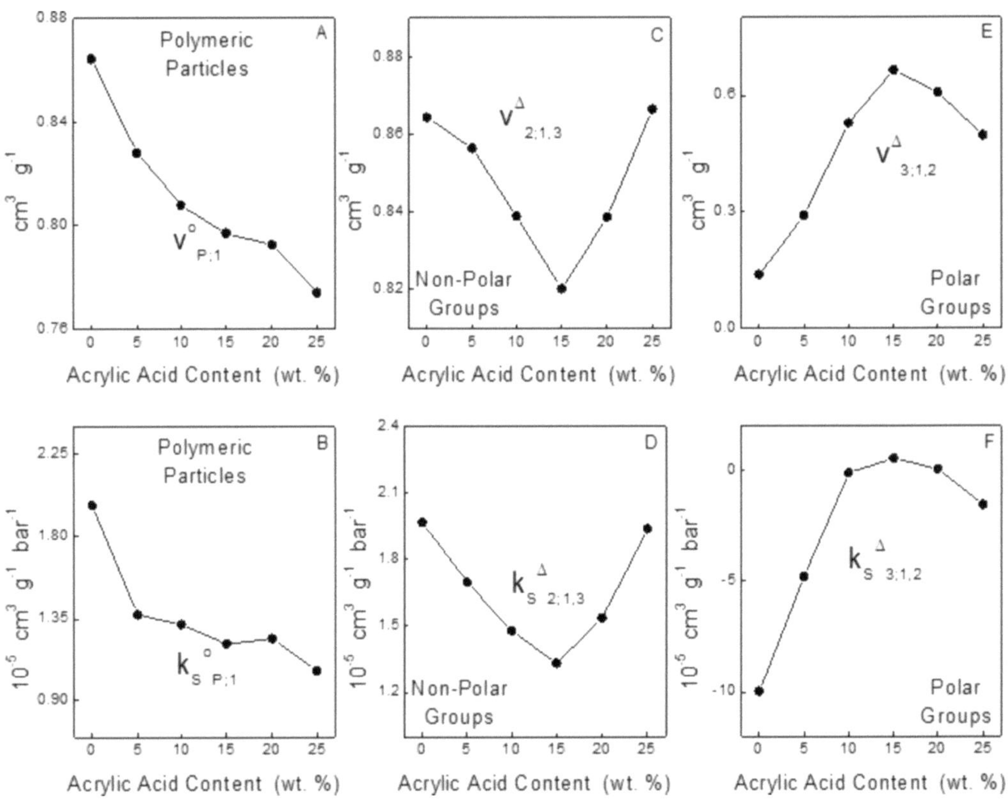

Figure 9. (A) Partial specific volume of the polymeric particles at infinite dilution as function of the polar group content. (B) Partial specific adiabatic compressibility of particles at infinite dilution as function of the polar group content. (C) Partial specific volume of non-polar groups at infinite dilution as function of the polar group content. (D) Partial specific adiabatic compressibility of non-polar groups at infinite dilution as function of the polar group content. (E) Partial specific volume of polar groups at infinite dilution as function of the polar group content. (F) Partial specific adiabatic compressibility of polar groups at infinite dilution as function of the polar group content.

where similar equations can be obtained for the adiabatic compressibilities. **Figure 9A** and **B** shows that $v^o_{P;1}$ and $k^o_{sP;1}$ decrease when the amount of polar groups increases. This fact indicates an increment of the hydration in the interior of the particle when the amount of polar groups increases. The distribution of this hydration is as follows. **Figure 9C** and **D** shows that $v^{\Delta}_{2;1,3}$ and $k^{\Delta}_{s2;1,3}$ decrease from 0 to 15% of polar groups, while **Figure 9E** and **F** shows that $v^{\Delta}_{3;1,2}$ and $k^{\Delta}_{s3;1,2}$ increase. This fact can be interpreted because the hydration is redistributed from the polar groups to the non-polar groups. In the region of 15–25%, this behaviour is reversed.

6. Conclusions

In this chapter, we have developed common thermodynamic bases for isothermal titration calorimetry, densimetry and measurement of sound speed in terms of thermodynamic partial properties (interaction partial enthalpies, partial volumes and partial adiabatic compressibilities). To build these common thermodynamic bases, it is necessary to introduce new concepts, i.e., the concept of fraction of a system and the concept of thermodynamic interaction between components of a system. An advantage of the proposed thermodynamic scheme is the possibility of including new thermodynamic partial properties as partial heat capacities.

Author details

Mónica Corea[1], Jean-Pierre E. Grolier[2] and José Manuel del Río[3]*

*Address all correspondence to: jm.delrio.garcia@gmail.com

1 National Polytechnic Institute, Mexico City, Mexico

2 Institut de Chimie de Clermont-Ferrand UMR 6296, Aubière Cedex, France

3 Petroleum Mexican Institute, Mexico City, Mexico

References

[1] do Carmo MP, editor. Differential Geometry of Curves and Surfaces. 2nd ed. Englewood Cliffs, New Jersey: Prentice-Hall, Inc.; 1976. p. 503

[2] Callen HB, editor. Thermodynamics and an Introduction to Thermostatics. 2nd ed. New York: John Wiley & Sons; 1985. p. 493

[3] Klotz IM, Rosenberg RM, editors. Chemical Thermodynamics: Basic Theory and Methods. 6th ed. New York: Wiley-Interscience; 2000. p. 556

[4] del Río JM, Santillan R, Vallejo I, Grolier J-PE, Jiménez-Ángeles F, Corea M. A thermodynamic method to study the interaction between NaOH and highly carboxylated polymeric particles in solution. Journal of Solution Chemistry. 2015;**44**:963–986

[5] Grolier J-PE, del Río JM. On the physical meaning of the isothermal titration calorimetry measurements in calorimeters with full cells. International Journal of Molecular Sciences. 2009;**10**:5296–5325

[6] Perez A, Ruíz A, Santillán R, del Río JM, Corea M. Partial volume and compressibility at infinite dilution. Reactive & Functional Polymers. 2008;**68**:1422–1428

[7] Santillán S, Nieves E, Alejandre R, Pérez E, del Río JM, Corea M. Comparative thermodynamic study of functional polymeric latex particles with different morphologies. Colloids and Surfaces A: Physicochemical and Engineering Aspects. 2014;**444**:189–208

[8] Chalikian TV, Sarvazyan AP, Breslauer KJ. Partial molar volumes, expansibilities, and compressibilities of alpha, omega amino carboxylic acids in aqueous solutions between 18 and 55°. Journal of Physical Chemistry. 1997;**97**:13017–13026

[9] Gekko K, Noguchi H. Compressibility of globular proteins in water at 25°C. Journal of Physical Chemistry. 1979;**83**:2706–2714

[10] Gekko K, Hasegawa Y. Compressibility—Structure relationship of globular proteins. Biochemistry. 1986;**25**:6563–6571

[11] Kauzmann W. Some factors in the interpretation of protein denaturation. Advances in Protein Chemistry. 1959;**14**:1–63

[12] Reisler E, Haik Y, Eisenberg H. Bovine serum albumin in aqueous guanidine hydrochloride solutions. Preferential and absolute interactions and comparison with other systems. Biochemistry. 1977;**16**:197–203

[13] Chalikian TV, Sarvazyan AP, Breslauer K. Hydration and partial compressibility of biological compounds. Journal of Biophysical Chemistry. 1994;**51**:89–109

[14] Clarke RG, Tremaine PR. Amino acids under hydrothermal conditions: Apparent molar volumes of aqueous alpha-alanine, beta-alanine, and proline at temperatures from 298 to 523 K and pressures up to 20.0 MPa. Journal of Physical Chemistry B. 1999;**103**:5131–5144

[15] Lo Surdo A, Shin C, Millero FJ. The apparent molal volume and adiabatic compressibility of some organic solutes in water at 25°C. Journal of Chemical & Engineering Data. 1978;**23**:197–201

[16] Gekko K, Yamagami K. Flexibility of food proteins as revealed by compressibility. Journal of Agricultural and Food Chemistry. 1991;**39**:57–62

[17] Chalikian TV, Totrov M, Abagyan R, Breslauer KJ. The hydration of globular-proteins as derived from volume and compressibility measurements—Cross correlating thermodynamic and structural data. Journal of Molecular Biology. 1996;**260**:588–603

[18] Gekko K, Kimoto A, Kamiyama T. Effects of disulfide bonds on compactness of protein molecules revealed by volume, compressibility, and expansibility changes during reduction. Biochemistry. 2003;**42**:13746–13753

[19] Dubins DN, Filfil R, Macgregor RB Jr, Chalikian TV. Role of water in protein-ligand interactions: Volumetric characterization of the binding of 2′-CMP and 3′CMP to ribonuclease A. Journal of Physical Chemistry B. 2000;**104**:390–401

[20] Grolier J-PE, del Rio JM. Isothermal titration calorimetry: A thermodynamic interpretation of measurements. Journal of Chemical Thermodynamics. 2012;**55**:193–202

[21] Grolier J-PE, del Rio JM. Isothermal titration calorimetry: Application of the Gibbs-Duhem equation to the study of the relationship between forward and reverse titrations. Journal of Solution Chemistry. 2014;**44**:987–1003

Section 3

Titrimetric Principles in Electrolytic Systems

Chapter 5

Principles of Titrimetric Analyses According to Generalized Approach to Electrolytic Systems (GATES)

Anna Maria Michałowska-Kaczmarczyk,
Aneta Spórna-Kucab and Tadeusz Michałowski

Additional information is available at the end of the chapter

http://dx.doi.org/10.5772/intechopen.69248

Abstract

The generalized equivalent mass (GEM) concept, based on firm algebraic foundations of the generalized approach to electrolytic systems (GATES), is considered and put against the equivalent "weight" concept, based on a "fragile" stoichiometric reaction notation still advocated by IUPAC. The GEM is formulated a priori, with no relevance to a stoichiometry. GEM is formulated in a unified manner, and referred to systems of any degree of complexity with special emphasis put on redox systems, where generalized electron balance (GEB) is involved. GEM is formulated on the basis of all attainable (and preselected) physicochemical knowledge on the system in question, and resolved with use of iterative computer programs. It is possible to calculate coordinates of the end points taken from the vicinity of equivalence point. This way, one can choose (among others) a proper indicator and the most appropriate (from analytical viewpoint) color change of the indicator. Some interpolative and extrapolative methods of equivalence volume V_{eq} determination are recalled and discussed. The GATES realized for GEM purposes provides the basis for optimization of analytical procedures a priori. The GATES procedure realized for GEM purposes enables to foresee and optimize new analytical methods, or modify, improve, and optimize old analytical methods.

Keywords: equilibrium analysis, mathematical modeling, redox titration curves, equivalence volume, Gran methods

1. Introductory remarks

Titrimetry reckons to the oldest analytical methods, still widely used because of high precision, accuracy, convenience, and affordability [1]. Nowadays, according to *Comité Consultatif pour la*

Quantité de la Matière (CCQM) opinion [2], it is considered as one of the primary methods of analysis i.e., it fulfills the demands of the highest metrological qualities. Titration is then perceived as a very simple and reliable technique, applied in different areas of chemical analysis. A physical chemist may perform a titration in order to determine equilibrium constants, whereas an analytical chemist performs a titration in order to determine the concentration of one or several components in a sample.

In a typical titration, V_0 mL of titrand (D) containing the analyte A of an unknown (in principle) concentration C_0 is titrated with V mL of titrant (T) containing the reagent B (C); V is the total volume of T added into D from the very beginning to a given point of the titration, where total volume of D + T mixture is $V_0 + V$, if the volume additivity condition is fulfilled. Symbolically, the titration T → D in such systems will be denoted as $B(C,V) \rightarrow A(C_0,V_0)$. Potentiometric acid-base pH titrations are usually carried out by using combined (glass + reference) electrode, responding to hydrogen-ion activity rather than hydrogen-ion concentration. Potentiometric titrations in redox systems are made with use of redox indicator electrodes (RIE) e.g., combined (Pt + reference) electrode [3–5]. For detection of specific ions in a mixture, ion-selective electrodes (ISE) are also used [5]. The degree of advancement of the reaction between B and A is the fraction titrated [6], named also as the degree of titration, and expressed as the quotient $\Phi = n_B/n_A$ of the numbers of mmoles: $n_B = C \cdot V$ of B and $n_A = C_0 \cdot V_0$ of A, i.e.,

$$\Phi = \frac{C \cdot V}{C_0 \cdot V_0} \tag{1}$$

We refer here to visual, pH, and potentiometric (E) titrations. The functional relationships between potential E or pH of a solution versus V or Φ, i.e., E = E(V) or E = E(Φ) and pH = pH(V) or pH = pH(Φ) functions, are expressed by continuous plots named as the related titration curves. The Φ provides a kind of normalization in visual presentation of the appropriate system. In the simplest case of acid-base systems, it is much easier to formulate the functional relationship Φ = Φ(pH), not pH = pH(Φ). In particular, the expression for Φ depends on the composition of D and T, see Appendix.

The detailed considerations in this chapter are based on principles of the generalized approach to electrolytic systems (GATES), formulated by Michałowski [9] and presented recently in a series of papers, related to redox [7–26] and nonredox systems [27–32] in aqueous and in mixed-solvent media [33–37]. The closed system separated from its environment by diathermal walls secure a heat exchange between the system and its environment, and realize dynamic processes in a quasistatic manner under isothermal conditions.

The mathematical description of electrolytic nonredox systems within GATES is based on general rules of charge and elements conservation. Nonredox systems are formulated with use of charge (ChB) and concentration balances $f(Y_g)$, for elements/cores $Y_g \neq$ H, O. The description of redox systems is complemented by generalized electron balance (GEB) concept, discovered by Michałowski as the Approach I to GEB (1992) and the Approach II to GEB (2006); GEB is considered as a law of a matter conservation, as the law of nature [7, 9, 11, 13, 25].

Formulation of redox systems according to GATES principles is denoted as GATES/GEB. Within the Approach II to GEB, based on linear combination $2{\cdot}f(O) - f(H)$ of the balances: $f(H)$ for H and $f(O)$ for O, the prior knowledge of oxidation degrees of all elements constituting the system is not needed; oxidants and reductants are not indicated. Moreover, the linear independency or dependency of $2{\cdot}f(O) - f(H)$ from other balances: ChB and $f(Y_g)$, is the general criterion distinguishing between redox and nonredox systems. Concentrations of the species within the balances are interrelated in a complete set of equations for equilibrium constants, formulated according to the mass action law principles. The GATES and GATES/GEB in particular, provide the best possible tool applicable for thermodynamic resolution of electrolytic systems of any degree of complexity, with the possibility of application of all physicochemical knowledge involved.

Several methods of equivalence volume (V_{eq}) determination are also presented in terms of the generalized equivalence mass (GEM) [8] concept, suggested by Michałowski (1979), with an emphasis put on the Gran methods and their modifications. The GEM concept has no relevance to a chemical reaction notation. Within GATES, the chemical reaction notation is only the basis to formulate the expression for the related equilibrium constant.

2. Formulation of generalized equivalent mass (GEM)

The main task of titration is the estimation of the equivalence volume, V_{eq}, corresponding to the volume $V = V_{eq}$ of T, where the fraction titrated (1) assumes the value

$$\Phi_{eq} = \frac{C \cdot V_{eq}}{C_0 \cdot V_0} \tag{2}$$

In contradistinction to visual titrations, where the end volume $V_e \cong V_{eq}$ is registered, all instrumental titrations aim, in principle, to obtain the V_{eq} value on the basis of experimental data $\{(V_j, y_j) \mid j = 1,\ldots,N\}$, where y = pH, E for potentiometric methods of analysis. Referring to Eq. (1), we have

$$C_0 \cdot V_0 = 10^3 \cdot m_A/M_A \tag{3}$$

where m_A [g] and M_A [g/mol] denote mass and molar mass of analyte (A), respectively. From Eqs. (1) and (3), we get

$$m_A = 10^{-3} \cdot C \cdot M_A \cdot V/\Phi \tag{4}$$

The value of the fraction V/Φ in Eq. (4), obtained from Eq. (1),

$$V/\Phi = C_0 \cdot V_0/C \tag{5}$$

is constant during the titration. Particularly, at the end (e) and equivalence (eq) points, we have

$$V/\Phi = V_e/\Phi_e = V_{eq}/\Phi_{eq} \tag{6}$$

The V_e [mL] value is the volume of T consumed up to the end (e) point, where the titration is terminated (ended). The V_e value is usually determined in visual titration, when a preassumed color (or color change) of D + T mixture is obtained. In a visual acid-base titration, pH_e value corresponds to the volume V_e (mL) of T added from the start for the titration and

$$\Phi_e = \frac{C \cdot V_e}{C_0 \cdot V_0} \tag{7}$$

is the Φ-value related to the end point. From Eqs. (4) and (6), one obtains:

$$m_A = 10^{-3} \cdot C \cdot V_e \cdot \frac{M_A}{\Phi_e} \tag{8a}$$

$$m_A = 10^{-3} \cdot C \cdot V_{eq} \cdot \frac{M_A}{\Phi_{eq}} \tag{8b}$$

This does not mean that we may choose between the two formulas: (8a) and (8b), to calculate m_A. Namely, Eq. (8a) cannot be applied for the evaluation of m_A: V_e is known, but Φ_e unknown; calculation of Φ_e needs prior knowledge of C_0 value; e.g., for the titration NaOH (C,V) → $HCl(C_0,V_0)$, see Appendix, we have

$$\Phi_e = \frac{C}{C_0} \times \frac{C_0 - \alpha_e}{C + \alpha_e} \text{ where } \alpha(\text{Appendix}), \text{and } \alpha_e = \alpha(pH_e) \tag{9}$$

However, C_0 is unknown before the titration; otherwise, the titration would be purposeless. The approximate pH_e value is known in visual titration. Also Eq. (8b) is useless: the "round" Φ_{eq} value is known exactly, but V_{eq} is unknown; V_e (not V_{eq}) is determined in visual titrations.

Because Eqs. (8a) and (8b) appear to be useless, the third, approximate formula for m_A, has to be applied, namely:

$$m_A' \cong 10^{-3} \cdot C \cdot V_e \cdot M_A/\Phi_{eq} = 10^{-3} \cdot C \cdot V_e \cdot R_A^{eq} \tag{10}$$

where Φ_{eq} is put for Φ_e in Eq. (8a), and

$$R_A^{eq} = \frac{M_A}{\Phi_{eq}} \tag{11}$$

is named as the equivalent mass. The relative error in accuracy, resulting from this substitution, equals to

$$\delta = (m_A' - m_A)/m_A = m_A'/m_A - 1 = V_e/V_{eq} - 1 = \Phi_e/\Phi_{eq} - 1 \tag{12}$$

For $\Phi_e = \Phi_{eq}$, we get $\delta = 0$ and $m_A' = m_A$; thus $\Phi_e \cong \Phi_{eq}$ (i.e., $V_e \cong V_{eq}$) corresponds to $m_A' \cong m_A$. A conscious choice of an indicator and a pH-range of its color change during the titration is possible on the basis of analysis of the related titration curve. From Eqs. (10) and (8b), we get

$$m_A = m_A'/(1+\delta) = m_A' \cdot \left(1 - \delta + \delta^2 - \ldots\right) \tag{13}$$

3. Accuracy and precision

In everyday conversation, the terms "accuracy" and "precision" are often used interchangeably, but in science—and analytical chemistry, in particular—they have very specific, and different definitions [38].

Accuracy refers to how close a result of measurement, e.g., expressed by concentration x (as an intensive variable), agrees with a known/true value x_0 of x in a sample tested. In N repeated trials made on this sample, we obtain x_j ($j = 1, \ldots, N$) and then the mean value $\overline{x}$ and variance s^2 are obtained

$$\overline{x} = \frac{1}{N} \cdot \sum_{j=1}^{N} x_j,\; s^2 = \frac{1}{N-1} \cdot \sum_{j=1}^{N} (x_j - x_0)^2 \tag{14}$$

The accuracy can be defined by the absolute value $|\overline{x} - x_0|$, whereas precision is defined by standard deviation, $s = (s^2)^{1/2}$; the accuracy and precision are brought here into the same units.

Accuracy and precision are the terms of (nearly) equal importance (weights: 1 and $(1 - 1/N)$ for the weighted sum of squares [39]) when involved in the relation [40, 41]

$$\frac{1}{N} \cdot \sum_{j=1}^{N} (x_j - x_0)^2 = 1 \cdot (x - x_0)^2 + (1 - 1/N) \cdot s^2 \tag{15}$$

where x_j — experimental ($j = 1, \ldots, N$) and true (x_0) values for x, $\overline{x}$ — mean value, s^2 — variance. The problem referred to accuracy and precision of different methods of V_{eq} determination has been raised, e.g., in Refs. [42, 43].

Accuracy and precision of the results obtained from titrimetric analyses depend both on a nature of D + T system considered and the method of V_{eq} evaluation. Herein, the kinetics of chemical reactions and transportation phenomena are of paramount importance.

4. The E = E(Φ) and/or pH = pH(Φ) functions

Relatively simple, functional relationships for $\Phi = \Phi(\mathrm{pH})$, ascribed to acid-base D + T systems, are specified in an elegant/compact form in Refs. [6, 27, 28, 30], see Appendix.

In acid-base systems occurred in aqueous media, pH is a monotonic function of V or Φ. From the relation,

$$\frac{dpH}{d\Phi} = \frac{dpH}{dV} \cdot \frac{dV}{d\Phi} = \frac{C_0 \cdot V_0}{C} \cdot \frac{dpH}{dV} \tag{16}$$

it results that the $\Phi = \Phi(pH)$ and $pH = pH(\Phi)$ relationships are mutually interchangeable, $C_0V_0/C > 0$. The relation (16) can be extended on other plots.

Explicit formulation of functional relationships: $\Phi = \Phi(pH)$ and $E = E(\Phi)$, is impossible in complex systems, where two or more different kinds (acid-base, redox, complexation, precipitation, liquid-liquid phase equilibria [44, 45]) of chemical reactions occur sequentially or/and simultaneously [8]. The E values are referred to SHE scale.

Monotonicity of $pH = pH(\Phi)$ and/or $E = E(\Phi)$ is not a general property in electrolytic redox systems. In **Figure 1**, the monotonic growth of $E = E(\Phi)$, i.e., $dE/d\Phi > 0$, is accompanied by monotonic growth of $pH = pH(\Phi)$, i.e., $dpH/d\Phi > 0$ [20].

In **Figure 2**, the monotonic drops of $E = E(\Phi)$, i.e., $dE/d\Phi < 0$, are accompanied by nonmonotonic changes of $pH = pH(\Phi)$ [9, 46, 47].

From inspection of **Figure 2B**, it results that the neighboring, *quasi* linear segments of the line (at $C_{Hg} = 0$) intersect at the equivalent points $\Phi_{eq1} = 2.5$ and $\Phi_{eq2} = 3.0$. So, it might seem that the pH titration is an alternative to the potentiometric titration method for the V_{eq} detection. It should be noted, however, that there are small changes within the pH range, where the characteristics of glass electrode is nonlinear, and an extended calibration procedure of this electrode is required. The opportunities arising from potential E measurement are here incomparably higher, so the choice of potentiometric titration is obvious.

In **Figure 3**, the nonmonotonic changes of $E = E(V)$ are accompanied by nonmonotonic changes of $pH = pH(V)$ [16].

The unusual shape of the respective plots for $E = E(\Phi)$ and $pH = pH(\Phi)$ is shown in **Figure 4** [13].

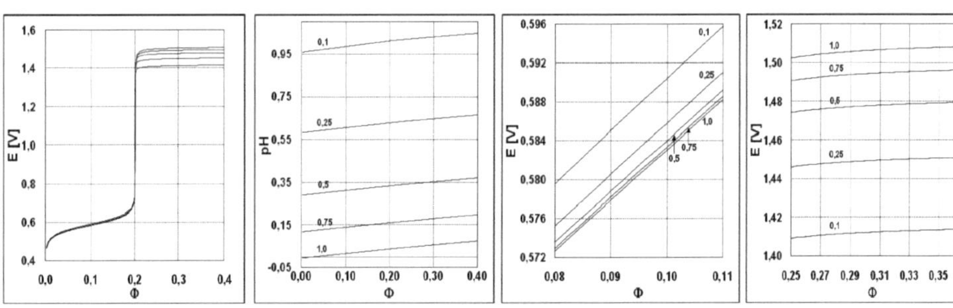

(A) (B) (C) (D)

Figure 1. The collected (A) $E = E(\Phi)$ and (B) $pH = pH(\Phi)$ curves plotted for D + T system $KMnO_4$ (C) → $FeSO_4$ (C_0) + H_2SO_4 (C_{01}) at $V_0 = 100$, $C_0 = 0.01$, $C = 0.02$, and different C_{01} values, indicated in Figures (B), (C), and (D) (in enlarged scales), before and after $\Phi = \Phi_{eq} = 0.2$.

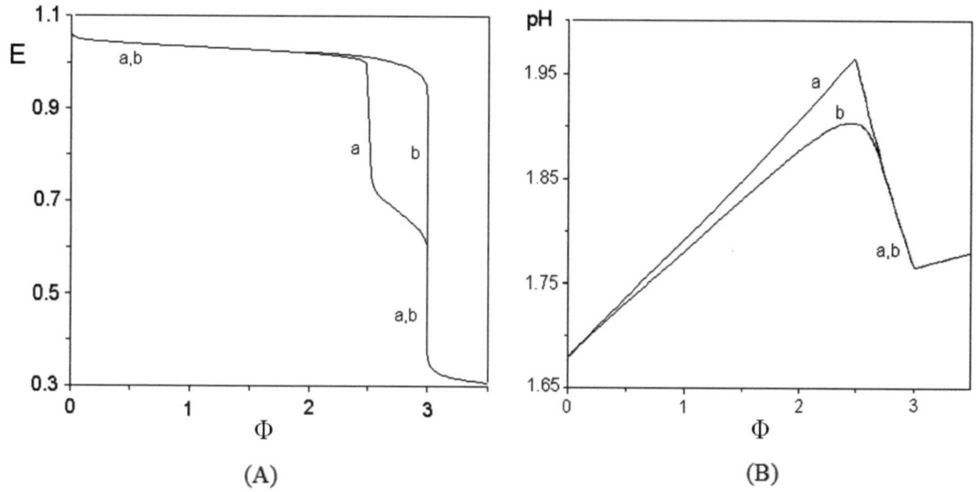

Figure 2. The theoretical plots of (A) E = E(Φ) and (B) pH = pH(Φ) functions for the D + T system, with KIO_3 ($C_0 = 0.01$) + HCl ($C_{01} = 0.02$) + H_2SeO_3 ($C_{Se} = 0.02$) + $HgCl_2$ (C_{Hg}) as D, and ascorbic acid $C_6H_8O_6$ (C = 0.1) as T; $V_0 = 100$, and (a) $C_{Hg} = 0$, (b) $C_{Hg} = 0.07$.

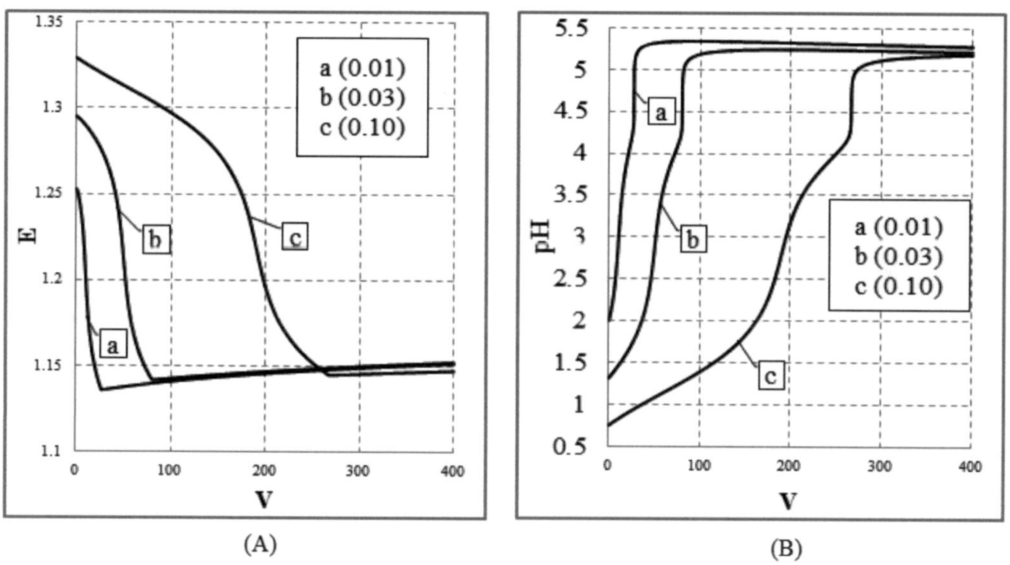

Figure 3. The theoretical plots of (A) E = E(V) and (B) pH = pH(V) functions for the system with $V_0 = 100$ mL of NaBr ($C_0 = 0.01$) + Cl_2 (C_{02}) as D titrated with V mL of $KBrO_3$ (C = 0.1) as T, at indicated (a, b, c) C_{02} values.

Other examples of the nonmonotonicity were presented in Refs. [7, 9, 46–49]. The nonmonotonic pH versus V relationships were also stated in experimental pH titrations made in some binary-solvent media [33]. Then, the Gran's statement "all titration curves are monotonic" [50] is not true, in general.

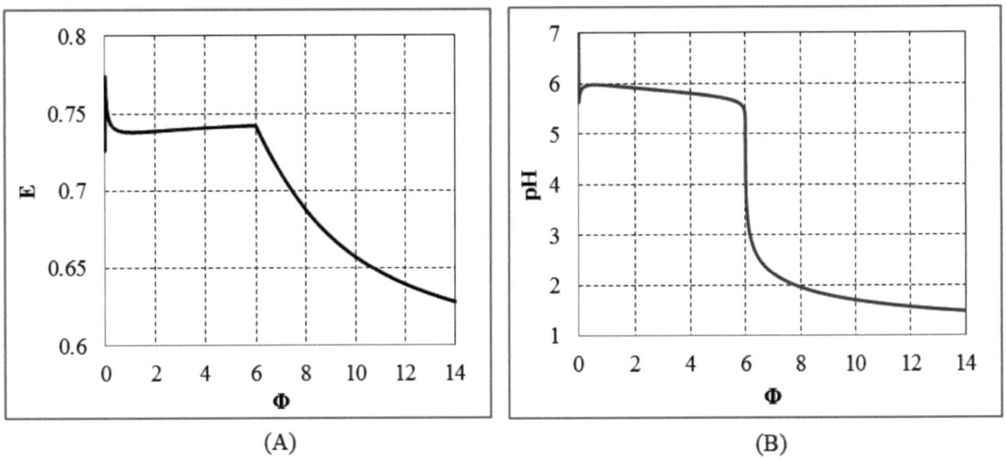

Figure 4. The plots of (A) $E = E(\Phi)$ and (B) $pH = pH(\Phi)$ functions for the system HI ($C = 0.1$) → KIO_3 ($C_0 = 0.01$).

5. Location of inflection and equivalence points

Some of the $E = E(\Phi)$ and/or $pH = pH(\Phi)$ (or $E = E(V)$ and/or $pH = pH(V)$) functions have inflection point(s), and characteristic S-shape (or reverse S-shape) is assumed within defined Φ (or V) range [51].

Generalizing, let us introduce the functions $y = y(V)$, where $y = E$ or pH and denote $V = V_{IP}$, with the volume referred to inflection point (IP) [52, 53], i.e., the point (V_{IP}, y_{IP}) of maximal slope $|\eta|$

$$\eta = \frac{dy}{dV} = \frac{1}{dV/dy} \tag{17}$$

on the related curve $y = y(V)$ ($y = E$, pH), plotted in normal coordinates (V, y) or their derivatives: $dy/dV = y_1(V)$ and $d^2y/dV^2 = y_2(V)$ on the ordinate. We have, by turns [54],

$$\frac{d^2y}{dV^2} = -\frac{1}{(dV/dy)^3} \cdot \frac{d^2V}{dy^2} \tag{18a}$$

$$\frac{d^2y}{dV^2} + \eta^3 \cdot \frac{d^2V}{dy^2} = 0 \tag{18b}$$

At $\eta \neq 0$, from Eq. (18b), we get $d^2V/dy^2 = 0$. Analogously to Eq. (16), we have

$$\frac{dE}{d\Phi} = \frac{C_0 \cdot V_0}{C} \cdot \frac{dE}{dV}$$

At the inflection point on the curve $y = y(V)$, we have maxima for $dy/d\Phi$ and $d^2y/dV^2 = 0$, see **Figure 5** for $y = E$ [55].

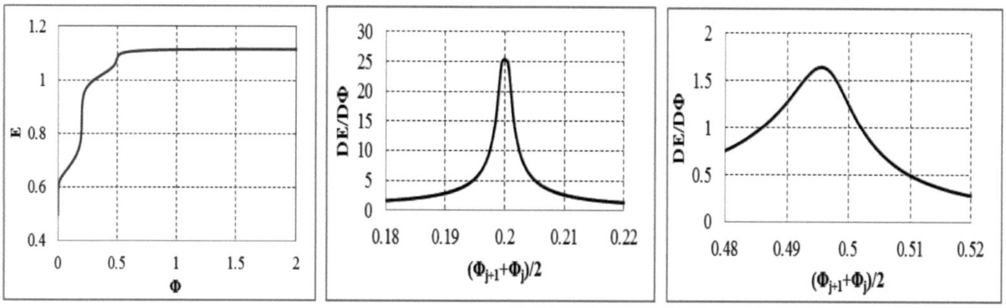

(A) (B) (C)

Figure 5. The function (A) $E = E(\Phi)$ and the difference quotient $DE/D\Phi = (E_{j+1} - E_j)/(\Phi_{j+1} - \Phi_j)$ versus $(\Phi_{j+1} + \Phi_j)/2$ relationships in the vicinity of $\Phi = 0.2$ (B) and $\Phi = 0.5$ (C) plotted for the system KIO_3 ($C = 0.1$) -> KI ($C_0 = 0.01$) + HCl ($C_{01} = 0.2$).

Referring to examples presented in **Figures 1A** and **2A**, we see that the inflection points (Φ_{IP}, E_{IP}) have the abscissas close to the related equivalence points (Φ_{eq}, E_{eq}), namely:

(0.2, 1.034)—see **Table 1** and **Figure 1A;**

(2.5, 0.903), (3.0, 0.414)—see **Table 2** and the curve a in **Figure 2A;**

(3.0, 0.652)—see **Table 2** and the curve b in **Figure 2A;**

Then we can consider Φ_{eq} (Eq. (2)) as a ratio of small natural numbers: p and q, i.e.,

$$\Phi_{eq} = \frac{p}{q} \quad (p,\ q \in N) \tag{19}$$

e.g., $\Phi_{eq} = 1$ (=1/1) for titration in D + T system with A = HCl and B = NaOH (see Eq. (9)); $\Phi_{eq} = 1/5 = 0.2$ in **Figure 1A** (see **Table 1**); $\Phi_{eq} = 5/2 = 2.5$ or $\Phi_{eq} = 3/1 = 3$ in **Figure 2A** (see **Table 2**).

Φ	**E**
0.19800	0.701
0.19900	0.719
0.19980	0.761
0.19990	0.778
0.19998	0.820
0.20000	1.034
0.20002	1.323
0.20010	1.365
0.20020	1.382
0.20200	1.442

Table 1. The (Φ, E) values related to $C_{01} = 0$ and other data presented in legend for **Figure 1A**.

$C_{Hg} = 0$				$C_{Hg} = 0.07$	
Φ	E	**Φ**	*E*	**Φ**	*E*
2.45	1.004	2.95	0.632	2.95	0.97
2.475	1	2.975	0.62	2.975	0.96
2.49	0.995	2.99	0.607	2.99	0.947
2.492	0.994	2.992	0.604	2.992	0.944
2.494	0.992	2.994	0.6	2.994	0.94
2.496	0.989	2.996	0.595	2.996	0.935
2.498	0.983	2.998	0.586	2.998	0.926
2.5	0.903	3	0.414	3	0.652
2.502	0.809	3.002	0.38	3.002	0.379
2.504	0.791	3.004	0.371	3.004	0.371
2.506	0.781	3.006	0.365	3.006	0.365
2.508	0.774	3.008	0.362	3.008	0.362
2.51	0.768	3.01	0.359	3.01	0.359
2.525	0.744	3.03	0.345	3.03	0.345
2.55	0.727	3.06	0.336	3.06	0.336

Table 2. The (Φ, E) values related to the data presented in legend for **Figure 2A**.

As we see (Eq. 12), the Φ_e values are compared each time with the "round" $\Phi_{eq} = p/q$ value for Φ_e due to the fact that just Φ_{eq} is placed in the denominator of the expression for the equivalent mass, $R_A{}^{eq}$ (Eq. (11)).

The Φ_e values, presented in **Tables 1** and **2** refer—in any case—to the close vicinity of the Φ_{eq} value(s), see e.g. $\Phi_{eq1} = 2.5$ and $\Phi_{eq2} = 3.0$.

Then from **Figures 1A** and **2A**, it results that location of IP is an interpolative method and $V_{IP} \cong V_{eq}$ [56], but in practice, this assumption may appear to be a mere fiction, especially in context with accuracy of measurements.

6. The case of diluted solutions

The V_{eq} and V_{IP} do not overlap in the systems of diluted solutions. For titration of V_0 mL of HB (C_0) with V mL of MOH (C), we have [6, 57]

$$V_{eq} - V_{IP} = \frac{x_{IP}}{1 + x_{IP}} \cdot (C_0/C + 1) \cdot V_0 \tag{20}$$

where

$$x_{IP} = \frac{8K_W}{C^2} + \left(\frac{8K_W}{C^2}\right)^2 + \dots \quad (21)$$

and $K_W = [H^{+1}][OH^{-1}]$. Similar relationship occurs for $AgNO_3$ (C,V) → NaCl (C_0,V_0) system; in this case, the relations [57]: Eq. (20) and

$$x_{IP} = \frac{8K_{sp}}{C^2} + \left(\frac{8K_{sp}}{C^2}\right)^2 + \dots \quad (22)$$

where $K_{sp} = [Ag^{+1}][Cl^{-1}]$, are valid.

7. Some interpolative methods of V_{eq} determination

7.1. The Michałowski method

Two interpolative methods, not based on the IP location, were presented by Fortuin [58] and Michałowski [6, 57]. The Fortuin method is based on an nomogram; an extended form of Fortuin's nomogram was prepared by the author of Ref. [6]. The Michałowski and Fortuin methods are particularly applicable to NaOH (C,V) → HCl (C_0,V_0) and NaOH (C,V) → HCl (C_0,V_0) systems. However, the applicability of the Michałowski method is restricted to diluted D and T, where the Fortuin method is invalid. In the Michałowski method, V_{eq} is the real and positive root of the equation

$$(1-2a)\cdot V_{eq}^3 + (2-3a)\cdot V_0\cdot V_{eq}^2 + V_0^2\cdot V_{eq} - a\cdot V_0^3 = 0 \quad (23)$$

where

$$a = \frac{1}{3}\cdot\frac{3A_0 - 2V_0A_1 + V_0^2A_2}{A_0 - V_0A_1 + V_0^3A_3} \quad (24)$$

and A_0, A_1, A_2, A_3 are obtained from results $\{(V_j, E_j) \mid j = 1,\dots, N\}$ of potentiometric titration, after applying the least squares method (LSM) to the function

$$\left(1 + \frac{V}{V_0}\right)^3 \cdot E = \sum_{i=0}^{3} A_i \cdot V^i \quad (25)$$

A useful criterion of validity of the V_{eq} value are: pK = – log K (K = K_W or K_{sp}) and standard redox potential (E_0), calculated from the formulas [59]:

$$pK = \log\left(\frac{24}{C^2}\right) + \log\left(-\frac{a_3}{a_1}\right);\ E_0 = a_0 + \frac{RT}{2F}\cdot \ln 10\cdot pK \quad (26)$$

where

$$a_3 = \frac{V_0^3}{3V_{eq}} \cdot \frac{3A_0 - 2V_0A_1 + V_0^2A_2}{(V_0 + V_{eq})^2}; \; a_1 = \frac{3a_3V_{eq}}{2V_0^2} \cdot (V_0 - V_{eq}) + \frac{V_0}{2} \cdot \frac{3A_0 - A_2V_0^2}{V_0 + V_{eq}}; \; a_0 = V_0^3 \cdot A_3 + a_1 + a_3 \tag{27}$$

7.2. The Fenwick–Yan method

The Yan method [59] is based on Newton's interpolation formula

$$f(x) = f(x_0) + \sum_{i=1}^{n} f_i(x_i) \cdot \prod_{j=0}^{i-1} (x - x_j) \tag{28}$$

where

$$f_1(x_j) = \frac{f(x_j) - f(x_0)}{x_j - x_0} \quad \text{for } j = 1, 2, \ldots, n$$

$$f_i(x_j) = \frac{f_{i-1}(x_j) - f_{i-1}(x_{i-1})}{x_j - x_{i-1}} \quad \text{for } j = i, \ldots, n$$

and on the assumption that $V_{eq} \cong V_{IP}$. Putting $n = 3$ in Eq. (28) and setting $d^2f(x)/dx^2 = 0$ for IP, after rearranging the terms one obtains

$$x_{IP} = \frac{1}{3} \cdot \left(x_0 + x_1 + x_2 - \frac{f_2(x_2)}{f_3(x_3)} \right) \tag{29}$$

Let $x_j = V_{k+j}$, $f(x_j) = y_{k+j}$, $j = 0, 1, 2, 3$; $y = \text{pH}$ or E. According to Yan's suggestion, $x_{IP} \cong V_{eq}$. Then, on the basis of 4 experimental points (V_{k+j}, y_{k+j}) $(j = 0, 1, 2, 3)$ taken from the immediate vicinity of V_{eq}, we get

$$V_{eq} = \frac{1}{3} \cdot \left(V_k + V_{k+1} + V_{k+2} - \frac{f_2(V_{k+2})}{f_3(V_{k+3})} \right) \tag{30}$$

Volumes V_{k+j} of T added were chosen from the immediate vicinity of V_{eq}. The best results are obtained if $V_{k+1} < V_{eq} < V_{k+2}$. The error in accuracy may be significant if $V_k < V_{eq} < V_{k+1}$ or $V_{k+2} < V_{eq} < V_{k+3}$. Moreover, the following conditions are also necessary for obtaining the accurate results: (i) volume increments $V_{k+i+1} - V_{k+I}$ (ca. 0.1 mL) are small and rather equal and (ii) concentrations of reagent in T and analyte in D are similar.

When the titrant is added in equal volume increments ΔV in the vicinity of the equivalence point, then $V_{k+j} - V_{k+i} = (j - i) \cdot \Delta V$, and Eq. (30) assumes the form

$$V_{eq} = V_{k+1} + \frac{y_k - 2y_{k+1} + y_{k+2}}{y_k - 3y_{k+1} + 3y_{k+2} - y_{k+3}} \cdot \Delta V \tag{31}$$

identical with one obtained earlier by Fenwick [60] on the basis of the polynomial function

$$y = A_0 + A_1 \cdot V + A_2 \cdot V^2 + A_3 \cdot V^3 \tag{32}$$

(compare it with Eq. (25)). In Ref. [6], it was stated that a simple equation for $x \cong V_{eq}$ can be obtained after setting $n = 4$ in Eq. (28). Then one obtains the following equation

$$6f_4(V_{k+4}) \cdot V_{eq}{}^2 + 3\left(f_3(V_{k+3}) - \beta \cdot f_4(V_{k+4})\right) \cdot V_{eq} + f_2(V_{k+2}) - \sigma \cdot f_3(V_{k+3}) + \gamma \cdot f_4(V_{k+4}) = 0 \tag{33}$$

where the parameters:

$$\sigma = V_k + V_{k+1} + V_{k=2},\ \beta = \sigma + V_{k+3},\ \gamma = \sum_{i>j=0}^{3} V_{k+i} \cdot V_{k+j}$$

are obtained on the basis of 5 points $\{(V_{k+j}, y_{k+j}) \mid j=0,\ldots,4\}$ from the close vicinity of V_{eq}.

8. Standardization and titrimetric analyses

The amount of an analyte in titrimetric analysis is determined from the volume of a titrant T (standard or standardized solution) required to react completely with the analyte in D. Titrations are based on standardization and determination steps. During the standardization, the titrant T with unknown concentration C of the species B is added into titrand D containing the standard S (e.g., potassium hydrogen phthalate, borax) with mass the m_S (g) known accurately. In this context, different effects involved with accuracy of visual titrations will be discussed.

Discussion on the formula 12 in context with Eq. (15) will be preceded by detailed considerations, associated with (1°) selection of an indicator (pH_e), (2°) volume V_0 of titrand D, (3°) concentration C_{0In} of indicator in D, (4°) buffer effect, and (5°) drop error, being considered as a whole. These effects will be considered first in context with nonredox systems. One should also draw attention whether the indicator is present in D as the salt or in the acidic form [61]; e.g., methyl orange is in the form of sodium salt, $NaIn = C_{14}H_{14}N_3NaO_3S$, more soluble than $HIn = C_{14}H_{15}N_3O_3S$.

To explain the effects 1° and 2°, we consider first a simple example, where the primary standard sample S is taken as an analyte A, A = S.

Example 1. We consider first the titration of $n_S = 1$ mmole of potassium hydrogen phthalate KHL solution with $C = 0.1$ mol/L NaOH. The equation for the related titration curve

$$\Phi = \frac{C}{C_0} \cdot \frac{(1 - \bar{n}) \cdot C_0 - \alpha}{C + \alpha} \tag{34}$$

is valid here [62], where α is specified in Appendix,

$$\overline{n} = \frac{2 \cdot [H_2L] + [HL^{-1}]}{[H_2L] + [HL^{-1}] + [L^{-2}]} = \frac{2 \cdot 10^{7.68-2pH} + 10^{4.92-pH}}{10^{7.68-2pH} + 10^{4.92-pH} + 1} \tag{35}$$

and $C_0 = 1/V_0$ (V_0 in mL). The values for the corresponding equilibrium constants are: $pK_W = 14$ for H_2O (in α), and $pK_1 = 2.76$, $pK_2 = 4.92$ for phthalic acid (H_2L).

The $\Phi = \Phi_e$ values in **Table 3** are calculated from Eq. (34) at some particular pH_e values, which denote limiting pH-values of color change for phenol red (6.4 ÷ 8.0), phenolphthalein (8.0 ÷ 10.0), and thymolphthalein (9.3 ÷ 10.5). A (unfavorable) dilution effect, expressed by different V_0 values, is involved here in context with particular indicators; at $pH_e = 6.4$, the dilution effect is insignificant, but grows significantly at higher pH_e values e.g., 10.5. As we see, at $pH_e = 8.0$, the $\Phi = \Phi_e$ value is closest to 1, assumed as Φ_{eq} in this case. At $pH_e = 6.4$ and 10.5, the Φ_e values differ significantly from 1. At $V_0 = 100$ and phenolphthalein used as indicator, at first appearance of pink color (pH ≈ 8.0), from Eq. (34) we have $\Phi_e = 0.9993 \Rightarrow \delta = -0.07\%$. The dilution practically does not affect the results of NaOH standardization against potassium hydrogen phthalate if pH titration is applied and titration is terminated at $pH_e \approx 8.0$ (**Table 3**).

A properly chosen indicator is one of the components of the D + T system in visual titrations. As a component of D having acid-base properties, the indicator should be included in the related balances [6, 62, 63]. The indicator effect, involved with its concentration, is considered in Examples 2 and 3. Moreover, the buffer effect is considered in Example 3.

Example 2. The equation of the titration curve for titration of V_0 mL of D containing $n_S = 1$ mmole of borax in the presence of C_{0In} mol/l methyl red ($pK_{In} = 5.3$) as an indicator with $C = 0.1$ mol/L HCl as T, is as follows [49, 62]

$$\Phi = \frac{C}{C_{0S}} \cdot \frac{(4\overline{n} - 10) \cdot C_{0S} + (1 - \overline{m}) \cdot C_{0In} + \alpha}{C - \alpha} \tag{36}$$

where α (Appendix), $C_0 = C_{0S} = 1/V_0$, and

$$\overline{n} = \frac{3 \cdot [H_3BO_3] + 2 \cdot [H_2BO_3] + [HBO_3]}{[H_3BO_3] + [H_2BO_3] + [HBO_3] + [BO_3]} = \frac{3 \cdot 10^{35.78-3pH} + 2 \cdot 10^{26.54-2pH} + 10^{13.80-pH}}{10^{35.78-3pH} + 10^{26.54-2pH} + 10^{13.80-pH} + 1} \tag{37}$$

pH_e	Φ_e		
	$V_0 = 50$	$V_0 = 100$	$V_0 = 200$
6.4	0.9679	0.9679	0.9678
8.0	0.9992	0.9993	0.9994
9.3	1.0012	1.0022	1.0051
10.0	1.0060	1.0010	1.0260
10.5	1.0190	1.0349	1.0825

Table 3. The Φ_e values for different $pH = pH_e$, calculated from Eq. (34), at C_0 and C values assumed in *Example* 1.

http://dx.doi.org/10.5772/intechopen.69248

$$\overline{m} = \frac{[HIn]}{[HIn] + [In]} = \frac{1}{1 + 10^{pH-5.3}} \tag{38}$$

It should be noted that the solution obtained after introducing 1 mmole of borax into water is equivalent to the solution containing a mixture of 2 mmoles of H_3BO_3 and 2 mmoles of NaH_2BO_3; $Na_2B_4O_7 + 5H_2O = 2H_3BO_3 + 2NaH_2BO_3$, resulting from complete hydrolysis of borax [62]. The results of calculations are presented in **Table 4**.

In context with **Table 4**, we refer to the one-drop error. For this purpose, let us assume that the end point was not attained after addition of V′ mL of titrant T, and the analyst decided to add the next drop of volume ΔV mL of the T. If the end point is attained this time, i.e., $V_e = V' + \Delta V$, the uncertainty in the T volume equals ΔV. Assuming ΔV = 0.03 mL and applying Eq. (1), we have:

$\Phi' = C \cdot V'/(C_0 \cdot V_0)$, $\Phi_e = C \cdot V_e/(C_0 \cdot V_0)$ and then $\Delta\Phi = \Phi_e - \Phi' = C \cdot V_e/(C_0 \cdot V_0) - C \cdot V'/(C_0 \cdot V_0) = C \cdot \Delta V/(C_0 \cdot V_0)$. At V_0 = 100 mL, C_0 = 0.01 mol/L, C = 0.1 mol/L, and ΔV = 0.03 mL, we have

$$\Delta\Phi = C \cdot \Delta V/(C_0 \cdot V_0) = 0.003 \tag{39}$$

Taking the value Φ_e = 2.0048 in **Table 4**, which refers to V_0 = 100 mL, C_0 = 0.01 mol/L, C = 0.1 mol/L, $C_{0In} = 10^{-5}$ mol/L and pH_e = 4.4, we see that |2.0048 – 2| = 0.0048 > 0.003 i.e., the discrepancy between Φ_{eq} and Φ_e is greater than the one assumed for ΔΦ = 0.003; it corresponds to ca. 1.5 drop of the titrant. At pH_e = 6.2 and other data chosen as previously, we get |1.9973 – 2| = 0.0027 < 0.003 i.e., this uncertainty falls within one–drop error.

The indicator effect stated in **Table 4**, for V_0 = 100, C_0 = 0.01, C = 0.1 and pH_e = 4.4 equals in Φ-units: |2.0048 2.0047| = 0.0001 at C_{0In} – 10^{-5} or |2.0058 – 2.0047| = 0.0011, i.e., it appears to be insignificant in comparison to ΔΦ = 0.003, and can therefore be neglected.

pH_e	**Φ_e**			
	C_{0In}	**$V_0 = 50$**	**$V_0 = 100$**	**$V_0 = 200$**
4.4	0	2.0027	2.0047	2.0087
	10^{-5}	2.0028	2.0048	2.0089
	10^{-4}	2.0033	2.0058	2.0109
5.3	0	1.9999	2.0001	2.0006
	10^{-5}	2.0001	2.0006	2.0016
	10^{-4}	2.0024	2.0051	2.0106
6.2	0	1.9964	1.9964	1.9965
	10^{-5}	1.9968	1.9973	1.9983
	10^{-4}	2.0008	2.0053	2.0142

Table 4. The Φ_e values calculated from Eqs. (36) to (38) for different pH = pH_e, C_{0In} and V_0 (mL) values assumed in Example 2. The pH_e values are related to the pH-interval <4.4 ÷ 6.2> corresponding to the color change of methyl red (HIn).

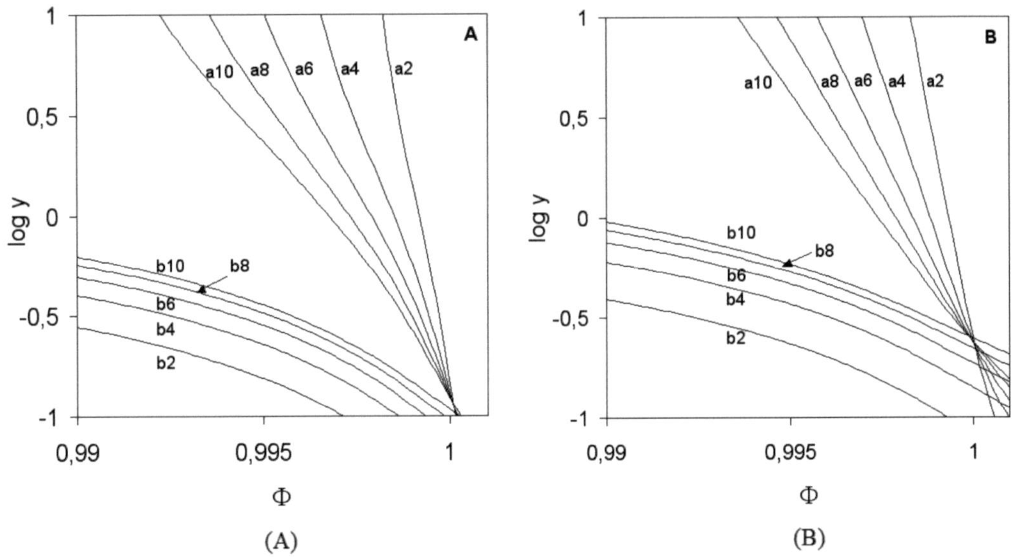

Figure 6. The logy versus Φ relationships in the close vicinity of $\Phi_{eq} = 1$, for $C_{In} = p \cdot 10^{-5}$ mol/L (p = 2, 4, 6, 8, 10); curves ap correspond to $C_{NH3} = 0.1$ mol/L, curves bp correspond to $C_{NH3} = 1.0$ mol/L; (A) refers to $r = 1$, (B) refers to $r = 4$.

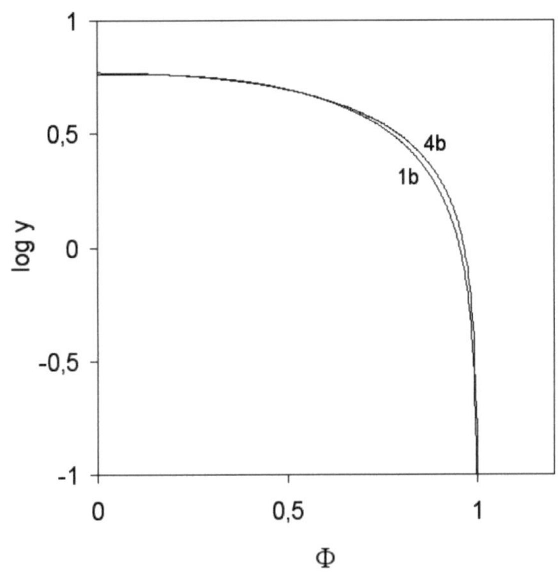

Figure 7. The logy versus Φ relationships plotted at $C_N = 1$ mol/L and $r = 1$ (curve 1b), and $r = 4$ (curve 4b).

Example 3. The solution of $ZnCl_2$ ($C_0 = 0.01$) buffered with NH_4Cl (C_1) and NH_3 (C_2), $C_1 + C_2 = C_N$, $r = C_2/C_1$, is titrated with EDTA (C = 0.02) in presence of Eriochrome Black T ($C_{In} = p \cdot 10^{-5}$, p = 2, 4, 6, 8, 10) as the indicator changes from wine red to blue color. The curves of logy versus Φ relationships, where

$$y = \frac{x_2}{x_1} \text{ and} : x_1 = \sum_{i=0}^{3} [H_iIn],\ x_2 = [ZnIn] + 2[ZnIn_2]$$

are plotted in **Figure 6**, where (A) refers to r = 1, (B) refers to r = 4. It is stated that at $C_N = 0.1$, the solution becomes violet (red + blue) in the nearest vicinity of $\Phi_{eq} = 1$, and the color change occurs at this point. At $C_N = 1.0$, the solution has the mixed color from the very beginning of the titration (**Figure 7**). At $C_N > 1.0$, the solution is blue from the start of the titration. This system was discussed in more details in Refs. [9, 37, 49, 62].

9. Intermediary comments

If a concentration C of the properly chosen reagent B in T is known accurately from the standardization, the B (C mol/L) solution can be used later as titrant T, applied for determination of the unknown mass m_A of the analyte A in D. The B (C) reacts selectively with an analyte A (C_0 mol/L) contained in the titrand (D). This way, NaOH is standardized as in Example 1, and HCl is standardized as in Example 2. In Example 3, the standard solution of EDTA can be prepared from accurately weighed portion of this preparation, without a need for standardization, if EDTA itself can be obtained in enough pure form.

The reaction between A and S, B and A, or S and A should be fast i.e., equilibrium is reached after each consecutive portion of T added in the titration made with use of calibrated measuring instrument and volumetric ware.

In pH or potentiometric (E) titration, the correct readout with use of the proper measuring instrument needs identical equilibrium conditions at the measuring electrode and in the bulk solution, after each consecutive portion of T added in a *quasistatic* a priori manner under isothermal conditions assumed in the D + T system.

The quasistaticity assumption is fulfilled only approximately; however, the resulting error in accuracy is affected by a drift involved with retardation of processes occurred at the indicator electrode against ones in the bulk solution, where titrant T is supplied. Then, the methods based on the inflection point (IP) registration give biased results, as a rule. This discrepancy can be limited to a certain degree, after slowing down the titrant dosage. Otherwise, the end point lags behind the equivalence point because of a slow response of the electrode.

In modern chemical analysis, titrations are performed automatically and the titrant is introduced continuously. In this context, the transportation factors concerning the response of the indicating system are of paramount importance. At low concentration of analyte, the degree of incompleteness of the reaction is the highest around the equivalence point, and then the methods based on the inflection point registration give biased results, as a rule. The results like ones obtained with precision 0.02% within 5 min of the potentiometric titration performed with use of an ion–selective electrode or alike (according to some literature reports), can be considered only as a mere fiction.

In this context, for the reasons specified above, it is safer to apply extrapolative methods of titrimetric analyses. Such a requirement is fulfilled by some methods applied in potentiometric analysis; the best known ones are the Gran methods considered e.g., in Refs. [3, 6, 65, 77]. The Gran methods of V_{eq} determination can replace the currently used first-derivative method in the potentiometric titration procedure.

In the mathematical model applied for V_{eq} evaluation, it is tacitly assumed that activity coefficients and electrode junction potentials are invariable during the titration. The slope of indicator electrode should be known accurately; the statement that the slope should necessarily be Nernstian [66] is not correct. In reference to acid-base titrations, T and D should not be contaminated by carbonate; it particularly refers to a strong base solution used as T [67, 68].

10. The Gran methods

10.1. Introductory remarks

The Gran methods is an eponym of the well–known methods of linearization of the S–shaped curves of potentiometric E or pH titration [69–71]. In principle, there are two original Gran methods, known as Gran I method (abbr. G(I)) [72] and Gran II (abbr. G(II)) method [73, 74].

In current laboratory practice, only G(II) is applied mainly in alkalinity [75] (referred to seawaters, as a rule) and acid–base titrations, in general. The presumable reasons of G(I) factual rejection (this statement was nowhere pronounced in literature) were clearly presented in the chapter [65], where G(I) and G(II) were thoroughly discussed. It was stated that the main reason of rejection was too high error, inherent in the simplified model that can be brought to the approximation

$$\ln(1+x) \cong x \tag{40}$$

to the first term of the related Maclaurin's series [76]

$$\ln(1+x) = \sum_{j=1}^{\infty} (-1)^{j+1} \cdot x^j/j$$

The relation Eq. (40) is valid only at $|x| << 1$. To extend the x range, Michałowski suggested the approximation [6]

$$\ln(1+x) = \frac{x}{1+x/2} \tag{41}$$

that appeared to be better than expansion of ln(1+x) into the Maclaurin series, up to the 18th term at $|x| \leq 1$ [65], see **Figure 8**.

It is noteworthy that some trials were done by Gran himself [50] to improve G(I), but his proposal based on some empirical formulas was a kind of "prosthesis" applied to the defective model. In further years, the name "Gran method" (in singular) has been factually limited to G(II) i.e., in literature the term "Gran method" is practically perceived as one tantamount with G(II).

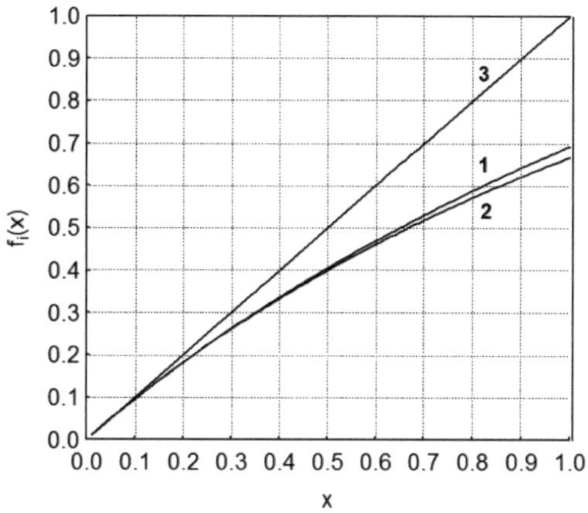

Figure 8. Comparison of the plots for: (1) $f_1(x) = \ln(1 + x)$, (2) $f_2(x) = x/(1 + x/2)$, and (3) $f_3(x) = x$ at different x-values, $0 < x \leq 1$.

10.2. The original Gran methods: G(I) and G(II)

The principle of the original Gran methods can be illustrated in a modified form [6], starting from titration of V_0 mL of C_0 mol/L HCl with V mL of C mol/L NaOH, taken as a simplest case. From charge and concentration balances, and $C_0V_0 = CV_{eq}$ i.e., $\Phi_{eq} = 1$ in Eq. (2), we get

$$([H^{+1}] - [OH^{-1}])(V_0 + V) = C \cdot (V_{eq} - V) \tag{42}$$

Applying the notations: $h = \gamma \cdot [H^{+1}]$, $ph = -\log h$, at $[H^{+1}] >> [OH^{-1}]$ (acid branch) i.e., $V < V_{eq}$, from Eq. (42) we have the relations:

$$(V_0 + V) \cdot 10^{-ph} = G_1 \cdot (V_{eq} - V) \tag{43}$$

$$ph \cdot \ln 10 = \ln(V_0 + V) - \ln G_1 + \ln(V_{eq} - V) \tag{44}$$

10.2.1. G(I) method

Applying Eq. (44) to the pair of points: (V_j, pH_j) and (V_{j+1}, pH_{j+1}), we have, by turns,

$$\ln 10 \cdot (pH_{j+1} - pH_j) = \ln\frac{V_0 + V_{j+1}}{V_0 + V_j} - \ln\frac{V_{eq} - V_{j+1}}{V_{eq} - V_j} \tag{45}$$

$$= \ln(1 + x_{1j}) - \ln(1 - x_{2j}) \tag{45a}$$

where:

$$x_{1j} = \frac{V_{j+1} - V_j}{V_0 + V_j} \tag{46a}$$

$$x_{2j} = \frac{V_{j+1} - V_j}{V_{eq} - V_j} \tag{46b}$$

Applying the approximation Eq. (40), we have:

$$\ln(1 + x_{1j}) \cong x_{1j};\ \ln(1 - x_{2j}) \cong -x_{2j} \tag{47}$$

Then we have, by turns,

$$\ln 10 \cdot (pH_{j+1} - pH_j) = x_{1j} + x_{2j} = (V_{j+1} - V_j) \cdot \frac{V_0 + V_{eq}}{(V_0 + V_j)(V_{eq} - V_j)} \tag{48}$$

$$y_j = G_1 \cdot (V_{eq} - V_j) + \varepsilon_j \tag{49}$$

$$y_j = P_1 - G_1 \cdot V_j + \varepsilon_j \tag{50}$$

where $P_1 = G_1 V_{eq}$, and

$$G_1 = \frac{\ln 10}{V_0 + V_{eq}} \tag{51}$$

$$y_j = \frac{1}{V_0 + V_j} \cdot \frac{V_{j+1} - V_j}{pH_{j+1} - pH_j} \tag{52}$$

From Eq. (50) and LSM, we get the formula

$$V_{eq} = \frac{P_1}{G_1} = \frac{\sum y_j V_j \cdot \sum V_j - \sum y_j \cdot \sum V_j^2}{N \cdot \sum y_j V_j - \sum y_j \cdot \sum V_j} \tag{53}$$

where $\sum = \sum_{j=1}^{N}$, and y_j is expressed by Eq. (52); it is the essence of G(I).

10.2.2. G(II) method

Eq. (43) can be rewritten into the regression equation

$$y_j = P_2 - G_2 \cdot V_j + \varepsilon_j \tag{54}$$

where:

$$G_2 = \gamma \cdot C \tag{55a}$$

$$P_2 = \gamma \cdot C \cdot V_{eq} = G_2 \cdot V_{eq} \tag{55b}$$

$$y_j = (V_0 + V_j) \cdot 10^{-ph_j} \tag{56}$$

Applying LSM to ph titration data {(V_j, ph_j) | j=1,…,N}, from (55b) we get

$$V_{eq} = \frac{P_2}{G_2} = \frac{\sum y_j V_j \cdot \sum V_j - \sum y_j \cdot \sum V_j^2}{N \cdot \sum y_j V_j - \sum y_j \cdot \sum V_j} \quad (57)$$

similar to Eq. (53), where y_j is expressed by Eq. (56) at this time; it is the essence of G(II).

10.3. The modified Gran methods

10.3.1. MG(I) method

Applying Eq. (41) to Eqs. (45a) and (46), we have

$$\ln(1 + x_{1j}) \cong \frac{x_{1j}}{1 + x_{1j}/2} = \frac{\frac{V_{j+1} - V_j}{V_0 + V_j}}{1 + \frac{V_{j+1} - V_j}{2(V_0 + V_j)}} = \frac{V_{j+1} - V_j}{V_0 + \frac{V_j + V_{j+1}}{2}} \quad (58a)$$

$$\ln(1 - x_{2j}) \cong \frac{-x_{2j}}{1 - x_{2j}/2} = \frac{-\frac{V_{j+1} - V_j}{V_{eq} - V_j}}{1 - \frac{V_{j+1} - V_j}{2(V_{eq} - V_j)}} = \frac{-(V_{j+1} - V_j)}{V_{eq} - \frac{V_j + V_{j+1}}{2}} \quad (58b)$$

From Eqs. (58) and (45a) we have, by turns,

$$\ln 10 \cdot (pH_{j+1} - pH_j) \cong (V_{j+1} - V_j) \cdot \left(\frac{1}{V_0 + V_j^*} + \frac{1}{V_{eq} - V_j^*} \right) = \frac{(V_{j+1} - V_j) \cdot (V_0 + V_{eq})}{(V_0 + V_j^*) \cdot (V_{eq} - V_j^*)} \quad (59)$$

$$y_j^* = G_1 \cdot (V_{eq} - V_j^*) + \varepsilon_j$$

$$y_j^* = P_1 - G_1 \cdot V_j^*) + \varepsilon_j \quad (60)$$

where G_1 and V_j^* are as in Eq. (51), and:

$$V_j^* = \frac{V_j + V_{j+1}}{2} \quad (61)$$

$$y_j^* = \frac{1}{V_0 + V_j^*} \cdot \frac{V_{j+1} - V_j}{pH_{j+1} - pH_j} \quad (62)$$

$$V_{eq} = \frac{P_1}{G_1} = \frac{\sum y_j^* V_j^* \cdot \sum V_j^* - \sum y_j^* \cdot \sum V_j^{*2}}{N \cdot \sum y_j^* V_j^* - \sum y_j^* \cdot \sum V_j^*} \quad (63)$$

Application of V_j^* in Eqs. (59) and (62), suggested in Ref. [6], improves the results of analyses when compared with Eqs. (50) and (52).

10.3.2. New algorithms referred to $Fe^{+2} + MnO_4^{-1}$ system

The algorithms applied below are referred to the system, where V_0 ml of the solution containing $FeSO_4$ (C_0) and H_2SO_4 (C_{01}) as D is titrated with V ml of $KMnO_4$ (C). The simplest form of GEB related to this system has the form [3, 46]

$$[Fe^{+2}] + [FeOH^{+1}] + [FeSO_4] - (5[MnO_4^{-1}] + 4[MnO_4^{-2}] + [Mn^{+3}] + [MnOH^{+2}])$$
$$= (C_0V_0 - 5CV)/(V_0 + V) = (1 - 5\Phi)C_0V_0/(V_0 + V) \quad (64)$$

Concentration balance for Fe has the form

$$[Fe^{+2}] + [FeOH^{+1}] + [FeSO_4] + [Fe^{+3}] + [FeOH^{+2}] + [Fe(OH)_2{}^{+1}] + 2[Fe_2(OH)_2{}^{+4}]$$
$$+ [FeSO_4{}^{+1}] + [Fe(SO_4)_2{}^{-1}] = C_0V_0/(V_0 + V) \quad (65)$$

On the basis of **Figure 9**, at $\Phi < \Phi_{eq} = 0.2$ and low pH-values, Eqs. (64) and (65) assume simpler forms:

$$[Fe^{+2}] + [FeSO_4] = (1 - 5\Phi) \times C_0V_0/(V_0 + V) \quad (66)$$

$$[Fe^{+2}] + [FeSO_4] + [Fe^{+3}] + [FeSO_4{}^{+1}] + [Fe(SO_4)_2^{-1}] = C_0V_0/(V_0 + V) \quad (67)$$

These simplifications are valid at low pH-values (**Figure 6**). Eqs. (66) and (67) can be rewritten as follows:

$$[Fe^{+2}] \cdot b_2 = (1–5\Phi)C_0V_0/(V_0 + V) \quad (68)$$

$$[Fe^{+2}] \cdot (b_2 + f_{23} \cdot b_3) = C_0 \cdot V_0/(V_0 + V) \quad (69)$$

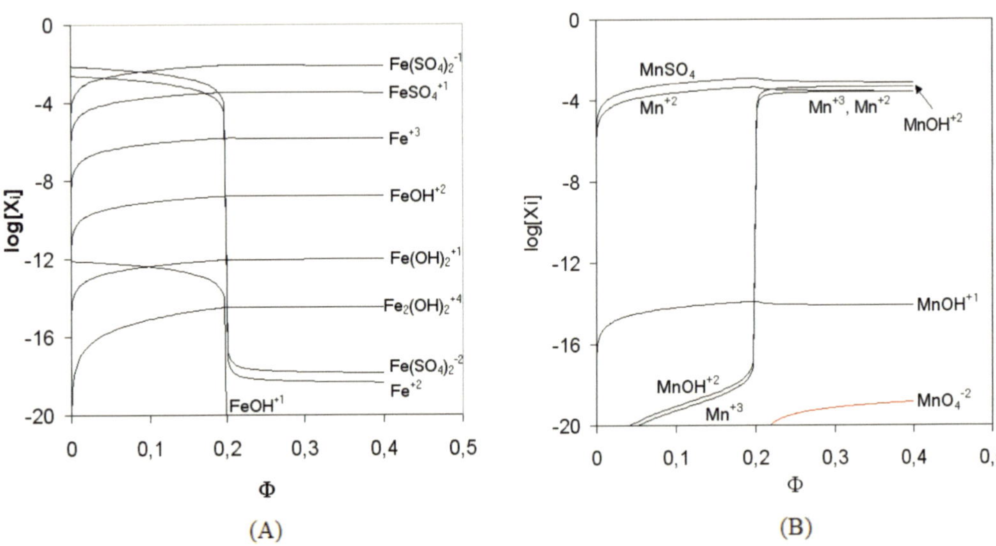

Figure 9. Dynamic speciation curves plotted for (A) Fe-species; (B) Mn-species in D + T system where $V_0 = 100$ mL of T ($FeSO_4$ ($C_0 = 0.01$) + H_2SO_4 ($C_{01} = 1.0$) is titrated with V ml of $KMnO_4$ ($C = 0.02$).

http://dx.doi.org/10.5772/intechopen.69248

valid for $\Phi < \Phi_{eq} = 0.2$, where:

$$b_2 = 1 + K_{21} \times [SO_4^{-2}] \tag{70a}$$

$$b_3 = 1 + K_{31} \times [SO_4^{-2}] + K_{32} \times [SO_4^{-2}]^2 \tag{70b}$$

$$f_{23} = \frac{[Fe^{+3}]}{[Fe^{+2}]} = 10^{A(E - E_0)} \tag{71a}$$

$$A = \frac{F}{R \cdot T \cdot \ln 10} = \frac{1}{a \cdot \ln 10} \tag{71b}$$

$$a = \frac{RT}{F} \tag{71c}$$

and $[FeSO_4] = K_{21}[Fe^{+2}][SO_4^{-2}]$, $[FeSO_4^{+1}] = K_{31}[Fe^{+3}][SO_4^{-2}]$, $[Fe(SO_4)_2^{-1}] = K_{32}[Fe^{+3}][SO_4^{-2}]^2$. From Eqs. (68) and (69), we have, by turns,

$$1 + f_{23} \cdot \frac{b_3}{b_2} = \frac{1}{1 - 5\Phi} \tag{72a}$$

$$10^{A(E - E_0)} \cdot \frac{b_3}{b_2} = \frac{5\Phi}{1 - 5\Phi} \tag{72b}$$

$$E = E_0 - a \cdot \ln\left(\frac{b_3}{b_2}\right) + a \cdot \ln(5\Phi) - a \cdot \ln(1 - 5\Phi) \tag{72c}$$

As results from **Figure 10**, the term $\ln(b_3/b_2)$ drops monotonically with Φ (and then V) value

$$\ln\left(\frac{b_3}{b_2}\right) = \alpha - \gamma \cdot \Phi \tag{73a}$$

$$\ln\left(\frac{b_3}{b_2}\right) = \alpha - \beta \cdot V \tag{73b}$$

The value for β in (73b) is small for higher C_{01} values, ca. 1 mol/L; in Ref. [77], it was stated that $\beta = 1.7 \cdot 10^{-3}$ at $C_{01} = 1.0$ mol/L; this change is small and can be neglected over the V-range covered in the titration. The assumption $\ln(b_3/b_2) =$ const is applied below in the simplified Gran models. For lower C_{01} values, this assumption provides a kind of drift introduced by the model applied, and then in accurate models, the formula Eq. (72c) is used.

From Eqs. (1) and (2), we have $\Phi/\Phi_{eq} = V/V_{eq}$; at $\Phi_{eq} = 0.2$, we get $5\Phi = V/V_{eq}$. Then applying Eq. (71b), we have

$$E = \omega - a \cdot (\alpha + \beta \cdot V) + a \cdot \ln\frac{V}{V_{eq}} - a \cdot \ln\left(1 - \frac{V}{V_{eq}}\right) \tag{74}$$

valid for $V < V_{eq}$, with the parameters: ω, α, β and a assumed constant within the V-range considered.

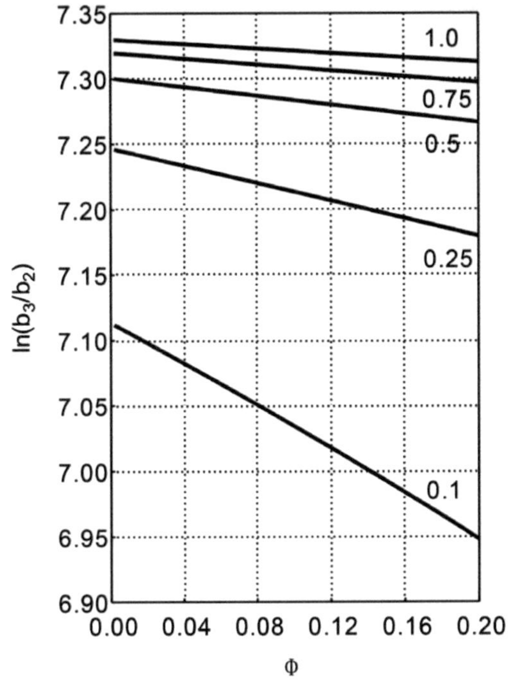

Figure 10. The $\ln(b_3/b_2)$ versus Φ relationships for the D + T system where $V_0 = 100$ mL of T ($FeSO_4$ ($C_0 = 0.01$) + H_2SO_4 (C_{01}) is titrated with V ml of $KMnO_4$ ($C = 0.02$). The lines are plotted at different concentrations (C_{01}) of H_2SO_4, indicated at the corresponding curves.

10.3.3. Simplified Gran I method

For jth and j+1th experimental point, from Eq. (72) we get:

$$E_j = E_0 - a \cdot \ln\frac{b_3}{b_2} + a \cdot \ln(5\Phi_j) - a \cdot \ln(1 - 5\Phi_j);$$

$$E_{j+1} = E_0 - a \cdot \ln\frac{b_3}{b_2} + a \cdot \ln(5\Phi_{j+1}) - a \cdot \ln(1 - 5\Phi_{j+1}) \quad (75)$$

$$E_{j+1} - E_j = a \cdot \ln\frac{\Phi_{j+1}}{\Phi_j} - a \cdot \ln\frac{1 - 5\Phi_{j+1}}{1 - 5\Phi_j}$$

Applying in Eq. (69) the identities: $\Phi_{j+1} = \Phi_j + \Phi_{j+1} - \Phi_j$ and $1 - 5\Phi_j = 1 - 5\Phi_{j+1} + 5(\Phi_{j+1} - \Phi_j)$ we have

$$E_{j+1} - E_j = a \cdot \ln(1 + x_{1j}) - a \cdot \ln(1 - x_{2j}) \quad (76)$$

where:

$$x_{1j} = (\Phi_{j+1} - \Phi_j)/\Phi_j \quad \text{and} \quad x_{2j} = 5(\Phi_{j+1} - \Phi_j)/(1 - 5\Phi_j) \quad (77)$$

http://dx.doi.org/10.5772/intechopen.69248

Applying the approximation Eq. (41) [6] for $x = x_{1j}$ and $x = -x_{2j}$ in Eq. (69) and putting $\Phi_j = C \cdot V_j/(C_0 \cdot V_0)$, $\Phi_{j+1} = C \cdot V_{j+1}/(C_0 \cdot V_0)$, we get, by turns,

$$\ln(1+x_{1j}) = \frac{\Phi_{j+1}-\Phi_j}{(\Phi_j+\Phi_{j+1})/2} = \frac{V_{j+1}-V_j}{V_j^*} \quad \text{and} \quad -\ln(1-x_{2j}) = \frac{5(\Phi_{j+1}-\Phi_j)}{1-5(\Phi_j+\Phi_{j+1})/2} = \frac{V_{j+1}-V_j}{V_{eq}-V_j^*} \tag{78}$$

$$\frac{1}{V_j^*} \cdot \frac{V_{j+1}-V_j}{E_{j+1}-E_j} = G_1 \cdot (V_{eq} - V_j^*) + \varepsilon_j \tag{79}$$

$$y_j^* = P_1 - G_1 \cdot V_j^* + \varepsilon_j \tag{80}$$

where V_j^* (Eq. (61)), and

$$y_j^* = \frac{1}{V_j^*} \cdot \frac{V_{j+1}-V_j}{E_{j+1}-E_j} \tag{81}$$

$$P_1 = \frac{1}{a}, G_1 = \frac{1}{a \cdot V_{eq}} \tag{82}$$

$$V_{eq} = \frac{P_1}{G_1} \tag{83}$$

P_1 and G_1 in Eq. (80) are obtained according to LSM, as previously described.

10.3.4. Accurate Gran I method

Applying analogous procedure based on Eqs. (67) and (68), we get, by turns,

$$E_{j+1} - E_j = a \cdot \gamma \cdot (\Phi_{j+1} - \Phi_j) + a \cdot \ln(1+x_{1j}) - a \cdot \ln(1-x_{2j}) \tag{84}$$

$$E_{j+1} - E_j = a \cdot \gamma \cdot (\Phi_{j+1} - \Phi_j) + a \cdot \frac{(\Phi_{j+1}-\Phi_j)}{(\Phi_{j+1}+\Phi_j)/2} + a \cdot \frac{5 \cdot (\Phi_{j+1}-\Phi_j)}{1-5(\Phi_{j+1}+\Phi_j)/2} \tag{85}$$

$$\frac{E_{j+1}-E_j}{V_{j+1}-V_j} = B + \frac{a}{V_j^*} + \frac{a}{V_{eq}-V_j^*} + \varepsilon_j \tag{86}$$

where

$$B = \frac{a \cdot \gamma}{5V_{eq}} \tag{87}$$

The parameters: B, a and V_{eq} are then found according to iterative procedure; V_j^* is defined by Eq. (61).

10.3.5. Simplified Gran II method

From Eqs. (1), (2) and (72a), we have, by turns

$$f_{23} \cdot \frac{b_3}{b_2} = \frac{\Phi}{\Phi_{eq} - \Phi} = \frac{V}{V_{eq} - V} \tag{88}$$

In this case, the fraction b_3/b_2 is assumed constant. From Eqs. (88) and (71a), we get, by turns,

$$V \cdot 10^{-A \cdot E} = \frac{b_3}{b_2} \cdot 10^{-A \cdot E_0} \cdot (V_{eq} - V) \tag{89}$$

If b_3/b_2 is assumed constant, then $G_2 = b_2/b_3 \cdot 10^{-A \cdot E_0} =$ const, and

$$V_j \cdot 10^{-A \cdot E} = P_2 - G_2 \cdot V_j + \varepsilon_j \tag{90}$$

Then

$$V_{eq} = \frac{P_2}{G_2} \tag{91}$$

where P_2 and G_2 are calculated according to LSM from the regression equation (90).

10.3.6. MG(II)A method

At $\beta \cdot V << 1$, we write

$$\frac{b_3}{b_2} = e^{\alpha} \cdot e^{-\beta V} \cong e^{\alpha} \cdot (1 - \beta \cdot V) \tag{92}$$

From Eqs. (89) and (92), we get

$$\Omega = \Omega(\vartheta, V) = V \cdot 10^{-E/\vartheta} = G_2 \cdot (V_{eq} - V) \cdot (1 - \beta \cdot V) \tag{93}$$

where $G_2 = e^{\alpha} \cdot 10^{-A \cdot E_0} =$ const and real slope ϑ of an electrode is involved, after putting $1/\vartheta$ for A. From Eq. (93), we have

$$\Omega = \Omega(\vartheta, V) = V \cdot 10^{-E/\vartheta} = P \cdot V^2 - Q \cdot V + R \tag{94}$$

where:

$$P = G_2 \cdot \beta \tag{95a}$$

$$Q = G_2 \cdot (\beta \cdot V_{eq} + 1) \tag{95b}$$

$$R = G_2 \times V_{eq} \tag{95c}$$

The P, Q, and R values in Eqs. (95a,b,c) are determined according to LSM, applied to the regression equation

$$\Omega_j = P \cdot V_j^2 - Q \cdot V_j + R + + \varepsilon_j \tag{96}$$

where

$$\Omega_j = V_j \cdot 10^{-E_j/\vartheta} \tag{97}$$

Then we get, by turns,

$$\frac{R}{P} = \frac{V_{eq}}{\beta}; \ \frac{Q}{R} = \beta + \frac{1}{V_{eq}}; P \cdot V_{eq}{}^2 - Q \cdot V_{eq} + R = 0 \tag{98}$$

$$V_{eq} = \frac{Q - \sqrt{Q^2 - 4 \cdot P \cdot R}}{2 \cdot P} \tag{99}$$

Eq. (96) is the basis for the modified G(II) method in its accurate version, denoted as MG(II)A method [77]. This method is especially advantageous in context of the error of analysis resulting from greater discrepancies $|\vartheta_c - \vartheta_p|$ between true (correct, ϑ_c) and preassumed (ϑ_p) slope values for RIE has been proved; the error in V_{eq} is significantly decreased even at greater $|\vartheta_c - \vartheta_p|$ values [77].

Numerous modifications of the Gran methods, designed also for calibration of redox indicator electrodes (RIE) purposes, were presented in the Refs. [4–6, 77]. Other calibration methods, related to ISE electrodes, are presented in Ref. [5].

10.4. Modified G(II) methods for carbonate alkalinity (CA) measurements

The G(II) methods were also suggested [28] and applied [78] for determination of carbonate alkalinity (CA) according to the modified CAM method. The CAM is related to the mixtures $NaHCO_3 + Na_2CO_3$ (system I) and $Na_2CO_3 + NaOH$ (system II), see **Table 5**. In addition to

No.	pH interval	Gran type functions	
		System I	System II
a	$pH > pK_2 + \Delta$	–	$(V_0 + V) \cdot 10^{ph} = C/K_W^* \cdot (V_a - V)$
b	$pK_2 - \Delta < pH \approx pK_2$	$(V_b + V) \cdot 10^{ph} = (K_2^*)^{-1} \cdot (V_c - V)$	$(V - V_a) \cdot 10^{ph} = (K_2^*)^{-1} \cdot (V_b - V)$
c	$pK_1 - \Delta \le pH \le pK_1 + \Delta$	$(V_d - V) \cdot 10^{-ph} = K_1^* \cdot (V - V_c)$	$(V_d - V) \cdot 10^{-ph} = K_1^* \cdot (V - V_c)$
d	$pH < pK_1 - \Delta$	$(V_0 + V) \cdot 10^{-ph} = \gamma \cdot C \cdot (V - V_d)$	$(V_0 + V) \cdot 10^{-ph} = \gamma \cdot C \cdot (V - V_d)$
Sequence of operations		d → c and b	d → c and b, a
Relationships		$V_d = V_{eq1} + V_{eq2}$	$V_d = V_{eq2} + V_{eq3}$
		$V_c = V_{eq2}/2$	$V_c = V_b = V_{eq2}/2 + V_{eq3}$
		$V_b = V_{eq1}$	$V_a = V_{eq3}$

Table 5. The modified Gran functions (CAM) related to the systems I and II (see text).

the determination of equivalence volumes, the proposed method gives the possibility of determining the activity coefficient of hydrogen ions (γ). Moreover, CAM can be used to calculate the dissociation constants (K_1, K_2) for carbonic acid and the ionic product of water (K_W) from a single pH titration curve. The parameters of the related functions are calculated according to LSM.

11. A brief review of other papers involved with titrimetric methods of analysis

11.1. Isohydric systems

Simple acid-acid systems are involved in isohydricity concept, formulated by Michalowski [31, 32, 79]. For the simplest case of acid-acid titration HB (C,V) → HL (C_0, V_0), where HB is a strong acid, HL is a weak monoprotic acid (K_1), the isohydricity condition, pH = const, occurs at

$$C_0 = C + C^2 \cdot 10^{pK_1} \qquad (100)$$

where $pK_1 = -\log K_1$.

In such a system, the ionic strength of the D + T mixture remains constant during the titration, i.e., the isohydricity and isomolarity conditions are fulfilled simultaneously and independently on the volume V of the titrant added. On this basis, a very sensitive method of pK_1 determination was suggested [31, 32]. The isohydricity conditions were also formulated for more complex acid-acid, base-base systems, etc.

11.2. pH titration in isomolar systems

The method of pH titration in isomolar D + T systems of concentrated solutions (ionic strength 2–2.5 mol/L) is involved with presence of equal volumes of the sample tested both in D and T. The presence of a strong acid HB in one of the solutions is compensated by a due excess of a salt MB in the second solution [80–90]. In the systems tested, acid-base and complexation equilibria were involved. The method enables to calculate concentrations of components in the sample tested together with equilibrium constants and activity coefficient of hydrogen ions. This method was applied for determination of a complete set of stability constants for mixed complexes [91–94].

11.3. Carbonate alkalinity, total alkalinity, and alkalinity with fulvic acids

Ref. [29] was referred to complex acid-base equilibria related to nonstoichiometric species involved with fulvic acids and their complexes with other metal ions and simpler species present in natural waters. For mathematical description of such systems, the idea of Simms constants was recalled from earlier issues e.g., Refs. [27, 28, 84–88], and the concept of activity/basicity centers in such systems was introduced.

11.4. Binary-solvent systems

Mutual pH titrations of weak acid solutions of the same concentration C in D and T formed in different solvents were applied [33–35] to formulate the $pK_i = pK_i(x)$ relationships for the acidity parameters, where x is the mole fraction of a cosolvent with higher molar mass in D + T mixture. The $pK_i = pK_i(x)$ relationship was based on the Ostwald's formula [95, 96] for monoprotic acid or the Henderson-Hasselbalch functions for diprotic and triprotic acids. The systems were modeled with the use of different nonlinear functions, namely Redlich-Kister and orthogonal (normal, shifted) Legendre polynomials. Asymmetric functions by Myers-Scott and the function suggested by Michałowski were also used for this purpose.

11.5. pH-static titration

Two kinds of reactions are necessary in V_{eq} registration according to pH–static titration; one of them has to be an acid–base reaction. The proton consumption or generation occurs in redox, complexation, or precipitation reactions [47], for example in titration of arsenite(+3) solution with I_2 + KI solution [18]; zinc salt solution with EDTA [97]; cyanide according to a (modified) Liebig-Denigès method [65, 102, 103].

11.6. Titration to a preset pH value

A cumulative effect of different factors on precision of V_{eq} determination was considered in [98] for pH titration of a weak monoprotic acids HL with a strong base, MOH. The results of calculations were presented graphically.

11.7. Dynamic buffer capacity

The dynamic buffer capacity concept, β_V, involving the dilution effect in acid-base D + T system, has been introduced [99] and extended in further papers [27, 28, 30, 100].

11.8. Other examples

The errors involved with more complex titrimetric analyses of chloride (mercurimetric method) [101], and cyanide (modified) Liebig-Denigès method) [97, 102, 103]. A modified, spectro-pH-metric method of dissociation constant determination was presented in Ref. [104]. An overview of potentiometric methods of titrimetric analyses was presented in Ref. [64]. The titration of ammonia in the final step of the Kjeldahl method of nitrogen determination [105, 106] was discussed in Ref. [107].

The proton consumption or generation occurs in redox, complexation, or precipitation reactions [47], for example in titration of arsenite(+3) solution with I_2 + KI solution [18]; zinc salt solution with EDTA [97]; cyanide according to a (modified) Liebig-Denigès method [65, 102, 103].

Three (complexation, acid-base, precipitation) kinds of reactions occur in the Liebig-Denigès method mentioned above. Four elementary (redox, complexation, acid-base, precipitation of I_2) types of reactions occur in the D + T system described in the legend for **Figure 2** and in less complex HCl → NaIO system presented in Ref. [21]. Other examples of high degree of complexity are shown in the works [9, 11, 12, 14–16]. One of the examples in Ref. [12] concerns a four-step analytical process with the four kinds of reactions, involving three electroactive elements.

12. Final comments

The Generalized Approach To Electrolytic Systems (GATES) provides the possibility of thermodynamic description of equilibrium and metastable, redox and non-redox, mono- and two-phase systems of any degree of complexity. It gives the possibility of all attainable/pre-selected physicochemical knowledge to be involved, with none simplifying assumptions done for calculation purposes. It can be applied for different types of reactions occurring in batch or dynamic systems, of any degree of complexity. The generalized electron balance (GEB) concept, discovered (1992, 2006) by Michałowski [11, 13] and obligatory for description of redox systems, is fully compatible with charge and concentration balance(s), and relations for the corresponding equilibrium constants.

The chapter provides some examples of dynamic electrolytic systems of different degree of complexity, realized in titrimetric procedure that may be considered from physicochemical and/or analytical viewpoints. In all instances, one can follow measurable quantities (potential E, pH) in dynamic and static processes, and gain the information about details not measurable in real experiments; it particularly refers to dynamic speciation. In the calculations made according to iterative computer programs, all physicochemical knowledge can be involved.

This chapter aims to demonstrate the huge/versatile possibilities inherent in GATES, as a relatively new quality of physicochemical knowledge gaining from electrolytic systems of different degrees of complexity, realizable with use of iterative computer programs.

Appendix

Expressions for Φ related to some D + T acid-base systems [6]; $M^{+1} = Na^{+1}, K^{+1}$; $B^{-1} = Cl^{-1}, NO_3^{-1}$; k = 0,…,n (nos. 1–10), k = 0,…,q − n (no. 11); l = 0,…,m.

No.	A	B	Φ =
1	HCl	MOH	$\frac{C}{C_0} \cdot \frac{C_0 - \alpha}{C + \alpha}$
2	MOH	HB	$\frac{C}{C_0} \cdot \frac{C_0 + \alpha}{C - \alpha}$
3	$M_kH_{n-k}L$	MOH	$\frac{C}{C_0} \cdot \frac{(n - k - \bar{n}) \cdot C_0 - \alpha}{C + \alpha}$
4	$M_kH_{n-k}L$	HB	$\frac{C}{C_0} \cdot \frac{(\bar{n} + k - n) \cdot C_0 + \alpha}{C - \alpha}$

http://dx.doi.org/10.5772/intechopen.69248

No.	A	B	$\Phi =$
5	$(NH_4)_kH_{n-k}L$	MOH	$\frac{C}{C_0}\cdot\frac{(n-k\cdot\bar{n}_N-\bar{n})\cdot C_0-\alpha}{C+\alpha}$
6	$(NH_4)_kH_{n-k}L$	HB	$\frac{C}{C_0}\cdot\frac{(\bar{n}+k\cdot\bar{n}_N-n)\cdot C_0+\alpha}{C-\alpha}$
7	$M_kH_{n-k}L$	$M_lH_{m-l}Ł$	$\frac{C}{C_0}\cdot\frac{(\bar{n}+k-n)C_0+\alpha}{(m-l-\bar{m})C-\alpha}$
8	$M_kH_{n-k}L$	$(NH_4)_lH_{m-l}Ł$	$\frac{C}{C_0}\cdot\frac{(\bar{n}+k-n)\cdot C_0+\alpha}{(m-l\cdot\bar{n}_N-\bar{m})\cdot C-\alpha}$
9	$(NH_4)_kH_{n-k}L$	$M_lH_{m-l}Ł$	$\frac{C}{C_0}\cdot\frac{(\bar{n}+k\cdot\bar{n}_N-n)\cdot C_0+\alpha}{(m-l-\bar{m})\cdot C-\alpha}$
10	$(NH_4)_kH_{n-k}L$	$(NH_4)_lH_{m-l}Ł$	$\frac{C}{C_0}\cdot\frac{(\bar{n}+k\cdot\bar{n}_N-n)\cdot C_0+\alpha}{(m-l\cdot\bar{n}_N-\bar{m})\cdot C-\alpha}$

The symbols:

$$\bar{n}=\frac{\sum_{i=1}^{q} i\cdot[H_iL^{+i-n}]}{\sum_{i=0}^{q}[H_iL^{+i-n}]}=\frac{\sum_{i=1}^{q} i\cdot 10^{\log K_{Li}^H-i\cdot pH}}{\sum_{i=0}^{q}10^{\log K_{Li}^H-i\cdot pH}}$$

$$\bar{m}=\frac{\sum_{i=1}^{p} i\cdot[H_iŁ^{+i-m}]}{\sum_{i=0}^{p}[H_iŁ^{+i-n}]}=\frac{\sum_{i=1}^{p} i\cdot 10^{\log K_{Łi}^H-i\cdot pH}}{\sum_{i=0}^{p}10^{\log K_{Łi}^H-i\cdot pH}}$$

$$\bar{n}_N=\frac{[NH_4^{+1}]}{[NH_4^{+1}]+[NH_3]}=\frac{10^{\log K_{1N}^H-pH}}{10^{\log K_{1N}^H-pH}+1}$$

enable to get a compact form of the functions, where:

$[H_iL^{+i-n}]=K_{Li}^H\cdot[H^+]^i[L^{-n}]$ $(i=0,\ldots,q)$; $[H_iŁ^{+i-m}]=K_{Łi}^H\cdot[H^+]^i[Ł^{-m}](i=0,\ldots,p)$; $[NH_4^{+1}]=K_{1N}^H[H^+][NH_3]$ $(\log K_{1N}^H=9.35)$; $K_{L0}^H=K_{Ł0}^H=1$; $M^{+1}=K^{+1}, Na^{+1}$; $[H^{+1}]=10^{-pH}$

and the ubiquitous symbol

$$\alpha=[H^{+1}]-[OH^{-1}]=10^{-pH}-10^{pH-pK_w}$$

termed as "proton excess" is used; $pK_W=14.0$ is assumed here.

Notations

D, titrand; T, titrant; V_0, volume of D; V, volume of T; all volumes are expressed in mL; all concentrations are expressed in mol/L.

Author details

Anna Maria Michałowska-Kaczmarczyk[1], Aneta Spórna-Kucab[2] and Tadeusz Michałowski[2*]

*Address all correspondence to: michalot@o2.pl

1 Department of Oncology, The University Hospital in Cracow, Cracow, Poland

2 Faculty of Chemical Engineering and Technology, Cracow University of Technology, Cracow, Poland

References

[1] Dasgupta, PK, Tanaka H, Jo KD. Continuous on-line true titrations by feedback based flow ratiometry. Application to potentiometric acid-base titrations. Analytica Chimica Acta. 2001;**435**:289–297

[2] Felber H, Rezzonico S, Mariassy M. Titrimetry at a metrological level. Metrologia **40** (2003) 249–254

[3] Michałowski T, Baterowicz A, Madej A, Kochana J. An extended Gran method and its applicability for simultaneous determination of Fe(II) and Fe(III). Analytica Chimica Acta. 2001;**442**(2):287–293

[4] Ponikvar M, Michałowski T, Kupiec K, Wybraniec S, Rymanowski M. Experimental verification of the modified Gran methods applicable to redox systems. Analytica Chimica Acta. 2008;**628**(2):181–189

[5] Michałowski T, Pilarski B, Ponikvar-Svet M, Asuero AG, Kukwa A, Młodzianowski J. New methods applicable for calibration of indicator electrodes. Talanta. 2011;**83**(5):1530–1537

[6] Michałowski T. Some remarks on acid-base titration curves. Chemia Analityczna. 1981; **26**:799–813

[7] Michałowski T. The Generalized Approach to Electrolytic Systems: I. Physicochemical and Analytical Implications. Critical Reviews in Analytical Chemistry. 2010;**40**(1):2–16

[8] Michałowski T, Pietrzyk A, Ponikvar-Svet M, Rymanowski M. The generalized approach to electrolytic systems: II. The generalized equivalent mass (GEM) concept. Critical Reviews in Analytical Chemistry. 2010;**40**(1):17–29

[9] Michałowski T. Application of GATES and MATLAB for resolution of equilibrium, metastable and non-equilibrium electrolytic systems, chapter 1 (pp. 1–34). In: Applications of MATLAB in Science and Engineering. Michałowski T, editors. Rijeka: InTech—Open Access publisher in the fields of Science, Technology and Medicine; 2011

[10] Michałowska-Kaczmarczyk AM, Asuero AG, Michałowski T. "Why not stoichiometry" versus "Stoichiometry – why not?" Part I. General context. Critical Reviews in Analytical Chemistry. 2015;**45**(2):166–188

[11] Michałowska-Kaczmarczyk AM, Asuero AG, Toporek M, Michałowski T. "Why not stoichiometry" versus "Stoichiometry – why not?" Part II. GATES in context with redox systems. Critical Reviews in Analytical Chemistry. 2015;**45**(3):240–268

[12] Michałowska-Kaczmarczyk AM, Michałowski T, Toporek M, Asuero AG. "Why not stoichiometry" versus "Stoichiometry – why not?" Part III, extension of GATES/GEB on complex dynamic redox systems. Critical Reviews in Analytical Chemistry. 2015;**45**(4): 348–366

[13] Michałowski T, Toporek M, Michałowska-Kaczmarczyk AM, Asuero AG. New trends in studies on electrolytic redox systems. Electrochimica Acta. 2013;**109**:519–531

[14] Michałowski T, Michałowska-Kaczmarczyk AM, Toporek M. Formulation of general criterion distinguishing between non-redox and redox systems, Electrochimica Acta. 2013;**112**:199–211

[15] Michałowska-Kaczmarczyk AM, Toporek M, Michałowski T. Speciation diagrams in dynamic iodide + dichromate system. Electrochimica Acta. 2015;**155**:217–227

[16] Toporek M, Michałowska-Kaczmarczyk AM, Michałowski T. Symproportionation versus disproportionation in bromine redox systems. Electrochimica Acta. 2015;**171**:176–187

[17] Michałowska-Kaczmarczyk AM, Michałowski T, Toporek M. Formulation of dynamic redox systems according to GATES/GEB principles. International Journal of Electrochemical Science. 2016;**11**:2560–2578

[18] Michałowski T, Ponikvar-Svet M, Asuero AG, Kupiec K. Thermodynamic and kinetic effects involved with pH titration of As(III) with iodine in a buffered malonate system. Journal of Solution Chemistry. 2012;**41**(3):436–446

[19] Michałowska-Kaczmarczyk AM, Michałowski T. Comparative balancing of non-redox and redox electrolytic systems and its consequences. American Journal of Analytical Chemistry. 2013;**4**(10):46–53

[20] Michałowska-Kaczmarczyk AM, Rymanowski M, Asuero AG, Toporek M, Michałowski T. Formulation of titration curves for some redox systems. American Journal of Analytical Chemistry. 2014;**5**:861–878

[21] Toporek M, Michałowska-Kaczmarczyk AM, Michałowski T. Disproportionation reactions of HIO and NaIO in static and dynamic systems. American Journal of Analytical Chemistry. 2014;**5**:1046–1056

[22] Michałowska-Kaczmarczyk AM, Michałowski T. Generalized electron balance for dynamic redox systems in mixed-solvent media. Journal of Analytical Sciences, Methods and Instrumentation. 2014;**4**(4):102–109

[23] Michałowska-Kaczmarczyk AM, Michałowski T. Compact formulation of redox systems according to GATES/GEB principles. Journal of Analytical Sciences, Methods and Instrumentation. 2014;**4**(2):39–45

[24] Michałowska-Kaczmarczyk AM, Michałowski T. GATES as the unique tool for simulation of electrolytic redox and non-redox systems. Journal of Analytical & Bioanalytical Techniques. 2014:**5**(4)

[25] Michałowski T. Generalized electron balance (GEB) as a law of preservation for electrolytic redox systems. The 65th Annual Meeting of the International Society of Electrochemistry, 31 August - 5 September, 2014, Lausanne, Switzerland, ise141331

[26] Meija J, Michałowska-Kaczmarczyk AM, Michałowski T. Redox titration challenge. Analytical and Bioanalytical Chemistry. 2017;**409**(1):11–13

[27] Asuero AG, Michałowski T. Comprehensive formulation of titration curves referred to complex acid-base systems and its analytical implications. Critical Reviews in Analytical Chemistry. 2011;**41**:151–187

[28] Michałowski T, Asuero AG. New approaches in modeling the carbonate alkalinity and total alkalinity. Critical Reviews in Analytical Chemistry. 2012;**42**:220–244

[29] Michałowska-Kaczmarczyk AM, Michałowski T. Application of Simms constants in modeling the titrimetric analyses of fulvic acids and their complexes with metal ions. Journal of Solution Chemistry. 2016;**45**:200–220

[30] Michałowska-Kaczmarczyk AM, Michałowski T. Dynamic buffer capacity in acid-base systems. Journal of Solution Chemistry. 2015;**44**:1256–1266

[31] Michałowski T, Pilarski B, Asuero AG, Dobkowska A. A new sensitive method of dissociation constants determination based on the isohydric solutions principle. Talanta. 2010;**82**(5):1965–1973

[32] Michałowski T, Pilarski B, Asuero AG, Dobkowska A, Wybraniec S. Determination of dissociation parameters of weak acids in different media according to the isohydric method. Talanta. 2011;**86**:447–451

[33] Pilarski B, Dobkowska A, Foks H, Michałowski T. Modeling of acid–base equilibria in binary-solvent systems: A comparative study. Talanta. 2010;**80**(3):1073–1080

[34] Michałowski T, Pilarski B, Dobkowska A, Młodzianowski J. Mathematical modeling and physicochemical studies on acid-base equilibria in binary-solvent systems. Wiadomości Chemiczne. 2010;**54**:124–154

[35] Michałowski T, Pilarski B, Asuero AG, Michałowska-Kaczmarczyk AM. Modeling of acid-base properties in binary-solvent systems, Chapter **9.4**. pp. 623–648. In: "Handbook of Solvents", Vol 1 Properties. Wypych G, editors. Toronto: ChemTec Publishing; 2014

[36] Michałowski T. Is the chemical equivalent due to the reaction equation notation? (in Polish). Orbital. 1999;**3/99**:174–176

[37] Michałowska-Kaczmarczyk AM, Michałowski T. Linear dependence of balances for non-redox electrolytic systems. American Journal of Analytical Chemistry. 2014;**5**:1285–1289

[38] https://en.wikipedia.org/wiki/Accuracy_and_precision

[39] https://en.wikipedia.org/wiki/Weight_function [accessed on August 5, 2017]

[40] Michałowski T, Rokosz A, Wójcik E. Optimization of the conventional method of determination of zinc as 8-oxyquinolinate in alkaline tartrate medium. Chemia Analityczna. 1980;**25**:563–566

[41] Michałowska-Kaczmarczyk M, Michałowski T. Simplex optimization and its applicability for solving analytical problems. Journal of Applied Mathematics and Physics. 2014;**2**:723–736

[42] Anfalt T, Jagner D. The precision and accuracy of some current methods for potentiometric end-point determination with reference to a computer-calculated titration curve. Analytica Chimica Acta. 1971;**57**:165–176

[43] Briggs TN, Stuehr JE. Simultaneous determination of precise equivalence points and pK values from potentiometric data. Single pK systems. Analytical Chemistry. 1974;**46**(11): 1517–1521

[44] Michałowski T, Lesiak A. Acid-base titration curves in disproportionating redox systems. Journal of Chemical Education. 1994;**71**(8):632–636

[45] Michałowski T. Effect of mutual solubility of solvents in multiple extraction. Journal of Chemical Education. 2002;**79**:1267–1268

[46] Michałowski T, Lesiak A. Formulation of generalized equations for redox titration curves. Chemia Analityczna (Warsaw). 1994;**39**:623–637

[47] Michałowski T, Rymanowski M, Pietrzyk A. Non-typical Brönsted's acids and bases. Journal of Chemical Education. 2005;**82**(3):470–472

[48] Michalowski T. Calculation of pH and potential E for bromine aqueous solutions. Journal of Chemical Education. 1994;**71**(7):560–562

[49] Michałowski T, N.Wajda, D.Janecki. A Unified quantitative approach to electrolytic systems. Chemia Analityczna (Warsaw). 1996;**41**:667–685

[50] Gran G. Equivalence volumes in potentiometric titrations. Analytica Chimica Acta. 1988;**206**:111–123

[51] Michałowska-Kaczmarczyk AM, Michałowski T, Asuero AG. Inflection points on some S-Shaped curves. Journal of Analytical Sciences, Methods and Instrumentation. 2014;**4**(1): 27–30

[52] Sigel H, Zuberbühler AD, Yamauchi O. Comments on potentiometric pH titrations and the relationship between pH-meter reading and hydrogen ion concentration. Analytica Chimica Acta. 1991;**255**(1):63–92

[53] http://www.titrations.info/acid-base-titration-polyprotics-and-mixtures [accessed on August 5, 2017]

[54] https://en.wikipedia.org/wiki/Inverse_functions_and_differentiation [accessed on August 5, 2017]

[55] https://en.wikipedia.org/wiki/Difference_quotient [accessed on August 5, 2017]

[56] Kolthoff IM, Furman NH. Potentiometric Titrations. 2nd ed. New York: Wiley; 1931

[57] Michałowski T, Stępak R. Evaluation of the equivalence point in potentiometric titrations with application to traces of chloride. Analytica Chimica Acta. 1985;**172**:207–214

[58] Fortuin JMH. Method for determination of the equivalence point in potentiometric titrations. Analytica Chimica Acta. 1961;**24**:175–191

[59] Yan JF. A method for the determination of equivalence point in potentiometric titrations using unequal volume increments. Analytical Chemistry. 1965;**37**(12):1588–1590

[60] Fenwick F. Rapid Method for Fixing End Point of Potentiometric Titration, Industrial and Engineering Chemistry. Analytical Edition. 1932;**4**(1):144–146

[61] https://en.wikipedia.org/wiki/PH_indicator [accessed on August 5, 2017]

[62] Michałowski T. Calculations in analytical chemistry with elements of computer programming (in Polish) PK, Cracow, 2001

[63] Michałowski T, Nizińska-Pstrusińska M, Sztark W, Baterowicz A. Laboratory trainings in analytical chemistry (in Polish). Wyd. PK, Kraków 2002

[64] Michałowski T. Potentiometric titration in acid-base systems; an overview of methods involved. Zeszyty Naukowe Uniwersytetu Jagiellońskiego. Prace Chemiczne. 1989;**32**:15–45

[65] Michałowski T, Toporek M, Rymanowski M. Overview on the Gran and other linearization methods applied in titrimetric analyses. Talanta. 2005;**65**(5):1241–1253

[66] Schwarz LM. Advances in acid-base Gran plot methodology. Journal of Chemical Education. 1987;**64**(11):947–950

[67] Michałowski T. Titration of monoprotic acids with sodium hydroxide contaminated by sodium carbonate. Journal of Chemical Education. 1988;**65**:181–182

[68] Michalowski T. Titrations of monoprotic acids with sodium hydroxide contaminated by sodium carbonate (the author replies). Journal of Chemical Education. 1990;**67**(12):1072

[69] Burns DT, Maitin BK, and Svehla G. Evaluation of equivalence points in the potentiometric titration of mixture of halides. Analyst. 1983;**108**:457

[70] Christian GD. Analytical Chemistry. 4th ed. Hong Kong: John Wiley & Sons; 1986, pp. 318–321

[71] Harris DG. Quantitative Chemical Analysis. 2nd ed. New York: W.H. Freeman & Co.; 1987. 243–244

[72] Gran G. Determination of the equivalent point in potentiometric titrations. Acta Chemica Scandinavica. 1950;**4**:559–577

[73] Gran G. Determination of the equivalence point in potentiometric titrations. Part II. Analyst. 1952;**77**:661–671

[74] Johansson A, Gran G. Automatic titration by stepwise addition of equal volumes of titrant. Part V. Extension of the Gran I method for calculation of the equivalence volume in acid-base titrations. Analyst. 1980;**105**:802–810

[75] Stumm W, Morgan JJ. Aquatic Chemistry: Chemical Equilibria and Rates in Natural Waters. 3rd ed. New York: John Wiley & Sons; 1996

[76] http://mathworld.wolfram.com/MaclaurinSeries.html [accessed on August 5, 2017]

[77] Michałowski T, Kupiec K, Rymanowski M. Numerical analysis of the Gran methods. A comparative study. Analytica Chimica Acta. 2008;**606**(2):172–183

[78] Pilarski B, Michałowska-Kaczmarczyk AM, Asuero AG, Dobkowska A, Lewandowska M, Michałowski T. A new approach to carbonate alkalinity. Journal of Analytical Sciences, Methods and Instrumentation. 2014;**4**:62–69

[79] Michałowski T, Asuero AG. Formulation of the system of isohydric solutions. Journal of Analytical Sciences, Methods and Instrumentation. 2012;**2**(1):1–4

[80] Michałowski T, Rokosz A, Tomsia A. Determination of basic impurities in mixture of hydrolysable salts. Analyst. 1987;**112**:1739–1741

[81] Michałowski T. Possibilities of application of some new algorithms for standardization purposes; Standardisation od sodium hydroxide solution against commercial potassium hydrogen phthalate. Analyst. 1988;**113**:833–935

[82] Michałowski T, Rokosz A, Negrusz-Szczęsna E. Use of Padé approximants in the processing of pH titration data; Determination of the parameters involved in the titration of acetic acid. Analyst. 1988;**113**:969–972

[83] Michałowski T, Rokosz A, Kościelniak P, Łagan J.M, J. Mrozek. Calculation of concentrations of hydrochloric and citric acids together in mixture with hydrolysable salts. Analyst. 1989;**114**:1689–1692

[84] Michałowski T. New Concepts of Analysis of Concentrated Solutions of Electrolytes with Use of Potentiometric Titration in Acid-Base Systems (in Polish), Kraków; 1989

[85] Michałowski T. Some remarks related to determination of free acids in presence of hydrolysable salts. Chemia Analityczna (Warsaw). 1990;**35**:375–390

[86] Michałowski T. Some new algorithms applicable to potentiometric titration in acid-base systems. Talanta. 1992;**39**(9):1127–1137

[87] Michałowski T, Gibas E. Applicability of new algorithms for determination of acids, bases, salts and their mixtures. Talanta. 1994;**41**(8):1311–1317

[88] Michałowski T, Rokosz A, Zachara M. Conventional method of determination of aminoacetic acid by pH-metric titration. Chemia Analityczna (Warsaw). 1994;**39**:217–222

[89] Michałowski T, Rokosz A, Gibas E. Standardization of sodium hydroxide against tartaric acid by pH-metric titration in isomolar systems. Acta Chimica Polonica. 1995;**37**:143–149

[90] Janecki D, Michałowski T, Zieliński M. A simple method of etidronate disodium determination in commercial preparations of the salt. Chemia Analityczna (Warsaw). 2000;**45**:659–666

[91] Janecki D, Doktór K, Michałowski T. Determination of stability constants of complexes of $M_iK_jH_kL$ type in concentrated solutions of mixed salts Talanta. 1999;**48**:1191–1197

[92] Janecki D, Doktór K, Michałowski T. Erratum to "Determination of stability constants of complexes of $M_jK_jH_kL$ type in concentrated solutions of mixed salts": [48 (1999) 1191] Talanta. 1999;**49**:943

[93] Janecki D, Michałowski T. Evaluation of equilibrium constants—a new approach. Chemia Analityczna (Warsaw). 1999;**44**:611–621

[94] Janecki D, Styszko-Grochowiak K, Michałowski T. The Catenation and Isomerisation effects on stability constants of complexes formed by some diprotic acids. Talanta. 2000;**52**:555–562

[95] Michałowska-Kaczmarczyk AM, Michałowski T. Ostwald's dilution law challenge. Analytical and Bioanalytical Chemistry. 2014;**406**(12):2741–2742

[96] Michałowska-Kaczmarczyk AM, Michałowski T. Solution to Ostwald's dilution law challenge. Analytical and Bioanalytical Chemistry. 2014;**406**(22):5251–5251

[97] Michałowski T, Toporek M, Rymanowski M. pH-Static titration: A quasistatic approach. Journal of Chemcal Education. 2007;**84**(1):142–150

[98] Michałowski T. On the precision of the method of titration to a preset pH value. Talanta. 1989;**36**(8):875–878

[99] Michałowski T, Parczewski A. A new definition of buffer capacity. Chemia Analityczna. 1978;**23**:959–964

[100] Michałowska-Kaczmarczyk AM, Michałowski T, Asuero AG. Formulation of dynamic buffer capacity for phytic acid. American Journal of Chemistry and Applications 2015;**2**(1): 5–9

[101] Kościelniak P, Michałowski T. An error of mercurimetric titration of chloride in presence od sodium nitroprusside as indicator. Chemia Analityczna. 1986;**31**:665–668

[102] Michałowski T. An error of Liebig-Deniges method of determination of the cyanide. Chemia Analityczna. 1983;**28**:313–315

[103] Michałowski T, Asuero AG, Ponikvar-Svet M, Toporek M, Pietrzyk A, Rymanowski M. Principles of computer programming applied to simulated pH–static titration of cyanide according to a modified Liebig-Denigès method. Journal of Solution Chemistry. 2012;**41**(7): 1224–1239

[104] Michałowski T., Wybraniec S., Ponikvar M. The modified method of spectro-pehametric determination of dissociation constants. Chemia Analityczna (Warsaw). 2008;**53**:737–742

[105] Sáez-Plaza P, Michałowski T, Navas MJ, Asuero AG, Wybraniec S. An overview of the Kjeldahl method of nitrogen determination. Part I. early history, chemistry of the procedure, and titrimetric finish. Critical Reviews in Analytical Chemistry. 2013;**43**(4):178–223

[106] Sáez-Plaza P, Navas MJ, Wybraniec S, Michalowski T, Asuero AG. An overview of Kjeldahl method of nitrogen determination. Part II. sample preparation, working scale, instrumental finish and quality control. Critical Reviews in Analytical Chemistry. 2013;**43**(4):224–272

[107] Michałowski T, Asuero AG, Wybraniec S. The titration in the Kjeldahl method of nitrogen determination: Base or acid as titrant?. Journal of Chemical Education. 2013;**90**(2): 191–197

Chapter 6

A Distinguishing Feature of the Balance 2•f(O)–f(H) in Electrolytic Systems: The Reference to Titrimetric Methods of Analysis

Anna Maria Michałowska-Kaczmarczyk,
Aneta Spórna-Kucab and Tadeusz Michałowski

Additional information is available at the end of the chapter

http://dx.doi.org/10.5772/intechopen.69249

Motto: 'Nothing is too wonderful to be true if it be consistent with the laws of Nature'
(M. Faraday)

Abstract

The balance $2{\cdot}f(O)-f(H)$ provides a general criterion distinguishing between electrolytic redox and non-redox systems of any degree of complexity, in aqueous, non-aqueous and mixed-solvent media. When referred to redox systems, it is an equation linearly independent on charge (ChB) and elemental/core balances $f(Y_g)$ for elements/cores $Y_g \neq H$ and O, whereas for non-redox systems, $2{\cdot}f(O)-f(H)$ is linearly dependent on these balances. The balance $2{\cdot}f(O)-f(H)$ formulated for redox systems is the primary form (pr-GEB) of the generalized electron balance (GEB) as the fundamental equation needed for resolution of these systems. Formulation of GEB for redox systems needs no prior knowledge of oxidation numbers for all elements of the system. Any prior knowledge of oxidation numbers for all elements in components forming a redox system and in the species of the system thus formed is not necessary within the Approach II to GEB. Oxidants and reductants are not indicated. Stoichiometry and equivalent mass are redundant concepts only. The GEB, together with charge balance and concentration balances for elements $\neq$ H and O, and the complete set of independent equations for equilibrium constants form an algorithm, resolvable with use of an iterative computer program. All attainable physicochemical knowledge can be included in the algorithm. Some variations involved with tests of possible reaction paths for metastable systems can also be made. The effects of incomplete physicochemical knowledge on the system can be also tested. One of the main purposes of this chapter is to provide the GEB formulation needed for resolution of redox systems and familiarize it to a wider community of chemists.

Keywords: electrolytes, redox systems, non-redox systems, generalized electron balance (GEB), titration

1. Introduction

Scientific theories describe particular units and rules governing the relationships between them. This description is considered as interpretation of reality. In particular, the thermodynamic description of any electrolytic system according to generalized approach to electrolytic systems (GATESs) [1] is based on fundamental and physical rules of conservation, expressed by charge balance (ChB) and elemental balances, for particular elements and/or cores in a closed system, separated from the environment by diathermal (freely permeable by heat) walls. The term 'core' is related to a cluster of elements with the same formula, structure and external charge. For example, $HSO_4^{-1}\cdot n_5H_2O$, $SO_4^{-2}\cdot n_6H_2O$ and $FeSO_4\cdot n_{27}H_2O$, in Eq. (3) (below), have the common core (SO_4^{-2}). In this context, the pairs of species: (i) $C_2O_4^{-2}$ and CO_3^{-2}; (ii) $C_2O_4^{-2}$ (from maleic acid) and $C_2O_4^{-2}$ (from fumaric acid) and (iii) NO_2^{-1} and NO_2 have no common cores.

Chemical interactions in electrolytic systems, for example protonation, neutralization, hydration, hydrolysis or dilution phenomena, are usually accompanied by exothermic or endothermic effects. However, the mass change, Δm, resulting from these thermal phenomena, estimated according to the Einstein's formula $\Delta E = \Delta m.c^2$ and put in context with their enthalpies ΔH ($\Delta E = \Delta H$), that is $\Delta m = \Delta H/c^2$, is negligibly small (not measurable). Therefore, the mass of a chemical system remains practically unchanged, regardless of whether the chemical reactions take place in it or not.

The heat exchange between the system and its environment through diathermal walls enables the temperature T of the system to be kept constant during the appropriate dynamic processes, such as titration, performed in a quasistatic manner. Stability of temperature T within the titrant (titrating solution, T), titrand (titrated solution, D) and D + T mixture, together with constancy of ionic strength (I) in D+T, is the preliminary condition ensuring stability of the corresponding equilibrium constants, $K_i = K_i(T, I)$, related to the system in question. The diathermal walls separate condensed (liquid or liquid + solid) phases from their environment.

An open chemical system is an approximation of the closed system—provided that the matter (e.g., H_2O, CO_2 and O_2) exchange between the system and its surroundings can be neglected, within a relatively short period of time needed to carry out the process considered, for example, titration.

On the initial stage of ChB and elemental/core balance formulation, it is advisable to start the quantitative considerations from the numbers of particular entities (components, species):

- N_{0j} for jth component constituting the system
- N_i for ith species in the system thus formed

For example, H_2O and gaseous HCl, as components, form an aqueous solution of HCl, with H_2O and hydrates of H^{+1}, OH^{-1} and Cl^{-1} as the species. Generally, when solid, liquid and/or gaseous solutes are introduced into water, a mono- or two-phase system is obtained. The resulting mixture is limited to the condensed (liquid or liquid + solid) phases. We refer mainly to aqueous media (W = H_2O), where the physicochemical knowledge is relatively extensive, incomparably better than that for the system with non-aqueous, or mixed-solvent media [2–6], with amphiprotic co-solvents involved, which is also considered in this chapter. For such media, the elemental $f(E_g)$) or core $f(core_g)$) balances written in terms of numbers of individual entities containing the elements (E_g) or cores ($core_g$), are formulated.

In aqueous electrolytic systems, different entities $X_i^{z_i}$ exist as hydrated species, $X_i^{z_i} \cdot n_iH_2O$; $n_i = n_{iW} = n_iH_2O$ is the mean number of water ($W = H_2O$) molecules attached to $X_i^{z_i}$ ($n_i \geq 0$), z_i is a charge of this species, expressed in elementary charge units and $e = F/N_A$ (F = Faraday constant, N_A = Avogadro number). For these species present in static or dynamic systems, we apply the notation

$$X_i^{z_i}(N_i, n_i) \tag{1}$$

where N_i is the number of these entities (individual species). On this basis, the numbers of particular elements in these species are calculated; for example, in the solution II (see below), N_{04} molecules of $FeSO_4{\cdot}7H_2O$ contain 14 N_{04} atoms of H, 11 N_{04} atoms of O and N_{11} atoms of Fe; N_5 ions of $HSO_4^{-1}{\cdot}n_5H_2O$ (N_5,n_5) in the set (2) of species specified below contain $N_5(1 + 2n_5)$ atoms of H, $N_5(4 + n_5)$ atoms of O and N_5 atoms of S.

In further parts of this chapter, the terms linear combination and linear dependency/independency of equations are introduced. These terms, well known from the elementary algebra course, will be applied to elemental/core balances, as a system of algebraic equations. The elemental balances $f(H)$ for hydrogen (H) and $f(O)$ for oxygen (O) and the linear combination $2{\cdot}f(O)-f(H)$ are formulated and then combined with charge balance (ChB) and other elemental/core balances $f(Y_g)$ for other elements ($Y_g = E_g$) or cores ($Y_g = core_g$), $Y_g \neq$ H and O. This way, the general properties of $2{\cdot}f(O)-f(H)$ in non-redox and redox systems are distinguished, see Refs. [7–18] and earlier references cited therein.

The $2f(O)-f(H)$, charge balance and elemental/core balances will be expressed first in terms of the numbers of particular entities. Next, the related balances will be presented in terms of molar concentrations, to be fully compatible with expressions for equilibrium constants that are also presented in terms of molar concentrations of the related species.

Static and dynamic systems are distinguished. A static system is obtained after disposable mixing with the respective components. For illustrative purposes, we consider first four solutions, as static non-redox systems, formed from the following components:

- (I) N_{01} molecules of $KMnO_4$, N_{02} molecules of CO_2 and N_{03} molecules of H_2O in V_1 mL of the resulting solution
- (II) N_{04} molecules of $FeSO_4{\cdot}7H_2O$, N_{05} molecules of H_2SO_4, N_{06} molecules of CO_2 and N_{07} molecules of H_2O in V_2 mL of the resulting solution
- (III) N_{08} molecules of $H_2C_2O_4{\cdot}2H_2O$, N_{09} molecules of H_2SO_4, N_{010} molecules of CO_2 and N_{011} molecules of H_2O in V_3 mL of the resulting solution
- (IV) N_{012} molecules of $FeSO_4{\cdot}7H_2O$, N_{013} molecules of $H_2C_2O_4{\cdot}2H_2O$, N_{014} molecules of H_2SO_4, N_{015} molecules of CO_2 and N_{016} molecules of H_2O in V_4 mL of the resulting solution.

The CO_2 in the respective solutions is primarily considered as one originated from ambient air, on the step of preparation of these solutions.

From these static systems, we prepare later different dynamic systems: (I) ⇒ (II), (I) ⇒ (III) and (I) ⇒ (IV), where (I) as titrant T is added into (II), (III) or (IV) as titrand D (T ⇒ D), and the D + T mixtures containing different species are formed.

To avoid a redundancy resulting from application of different subscripts within (N_i, n_i) ascribed to the same species $X_i^{z_i} \cdot n_iH_2O$ in different solutions (I)–(IV), we apply the common basis of the species from which the components will be selected to the respective balances. The set of the species is as follows:

$H_2O(N_1)$, $H^{+1}(N_2, n_2)$, $OH^{-1}(N_3, n_3)$, $K^{+1}(N_4, n_4)$, $HSO_4^{-1}(N_5, n_5)$, $SO_4^{-2}(N_6, n_6)$, $H_2C_2O_4(N_7, n_7)$, and $HC_2O_4^{-1}(N_8, n_8)$, $C_2O_4^{-2}(N_9, n_9)$, $H_2CO_3(N_{10}, n_{10})$, $HCO_3^{-1}(N_{11}, n_{11})$, $CO_3^{-2}(N_{12}, n_{12})$, $MnO_4^{-1}(N_{13}, n_{13})$, $MnO_4^{-2}(N_{14}, n_{14})$, $Mn^{+3}(N_{15}, n_{15})$, $MnOH^{+2}(N_{16}, n_{16})$, $MnC_2O_4^{+1}(N_{17}, n_{17})$, $Mn(C_2O_4)_2^{-1}(N_{18}, n_{18})$, $Mn(C_2O_4)_3^{-3}(N_{19}, n_{19})$, $Mn^{+2}(N_{20}, n_{20})$, $MnOH^{+1}(N_{21}, n_{21})$, $MnSO_4(N_{22}, n_{22})$, $MnC_2O_4(N_{23}, n_{23})$, $Mn(C_2O_4)_2^{-2}$ (N_{24}, n_{24}), $Fe^{+2}(N_{25}, n_{25})$, $FeOH^{+1}(N_{26}, n_{26})$ and $FeSO_4(N_{27}, n_{27})Fe(C_2O_4)_2^{-2}(N_{28}, n_{28})$, $Fe(C_2O_4)_3^{-4}(N_{29}, n_{29})$, $Fe^{+3}(N_{30}, n_{30})$, $FeOH^{+2}(N_{31}, n_{31})$, $Fe(OH)_2^{+1}(N_{32}, n_{32})$, $Fe_2(OH)_2^{+4}(N_{33}, n_{33})$, $FeSO_4^{+1}(N_{34}, n_{34})$, $Fe(SO_4)_2^{-1}(N_{35}, n_{35})$, $FeC_2O_4^{+1}(N_{36}, n_{36})$, $Fe(C_2O_4)_2^{-1}(N_{37}, n_{37})$, $Fe(C_2O_4)_3^{-3}(N_{38}, n_{38})$, and $\mathbf{FeC_2O_4}(N_{39}, n_{39})$, $\mathbf{MnC_2O_4}(N_{40}, n_{40})$ (2)

2. A short note

Referring to pure algebra, let us consider the set of G + 1 algebraic equations: $f_g(\mathbf{x}) = \varphi_g(\mathbf{x}) - b_g = 0$, where g = 0,1,…,G, $\mathbf{x}^T = (x_1,…,x_I)$, transposed (^{T}) vector $\mathbf{x}$, composed of independent (scalar) variables x_i (i ϵ <1, I>); a_{gi}, b_g ϵ $\mathcal{R}$ are independent (explicitly) on $\mathbf{x}$. After multiplying the equations by the numbers ω_g ϵ $\mathcal{R}$, and addition of the resulting equations, we get the linear combination $\sum_{g=0}^{G} \omega_g \cdot f_g(\mathbf{x}) = 0 \Leftrightarrow \sum_{g=0}^{G} \omega_g \cdot \varphi_g(\mathbf{x}) = \sum_{g=0}^{G} \omega_g \cdot b_g$ of the basic equations.

Formation of linear combinations is applicable to check the linear dependence or independence of the balances. A very useful/effective manner for checking/stating the linear dependence of the balances is the transformation of an appropriate system of equations to the identity 0 = 0 [2, 23]. For this purpose, we will try, in all instances, to obtain the simplest form of the linear combination. To facilitate these operations, carried out by cancellation of the terms on the left and right sides of equations after multiplying and changing sides of these equations, we apply the equivalent forms of the starting equations $f_g(\mathbf{x}) = 0$:

$$f_g(\mathbf{x}) : \varphi_g(\mathbf{x}) - b_g = 0 \Leftrightarrow \varphi_g(\mathbf{x}) = b_g \Leftrightarrow -f_g(\mathbf{x}) : -\varphi_g(\mathbf{x}) = -b_g \Leftrightarrow b_g = \varphi_g(\mathbf{x}). \quad (3)$$

In this notation, $f_g(\mathbf{x})$ will be essentially treated not as the algebraic expression on the left side of the equation $f_g(\mathbf{x}) = 0$ but as an equation that can be expressed in alternative forms presented above.

http://dx.doi.org/10.5772/intechopen.69249

3. Combination of elemental/core balances for non-redox systems

For the solution (I), we have the balances:

$$f_1 = f(\mathrm{H}):$$

$$2N_1 + N_2(1+2n_2) + N_3(1+2n_3) + 2N_4n_4 + N_{10}(2+2n_{10}) + N_{11}(1+2n_{11})$$

$$+ 2N_{12}n_{12} + 2N_{13}n_{13} = 2N_{03}$$

$$f_2 = f(\mathrm{O}):$$

$$N_1 + N_2n_2 + N_3(1+n_3) + N_4n_4 + N_{10}(3+n_{10}) + N_{11}(3+n_{11}) + N_{12}(3+n_{12})$$

$$+ N_{13}(4+n_{13}) = 4N_{01} + 2N_{02} + N_{03}$$

$$f_{12} = 2\cdot f(\mathrm{O}) - f(\mathrm{H}):$$

$$-N_2 + N_3 + 4N_{10} + 5N_{11} + 6N_{12} + 8N_{13} = 8N_{01} + 4N_{02}$$

$$f_0 = \mathrm{ChB}: \quad N_2 - N_3 + N_4 - N_{11} - 2N_{12} - N_{13} = 0$$

$$-f_3 = -f(\mathrm{K}): \quad N_{01} = N_4$$

$$-7f_4 = -7f(\mathrm{Mn}): \quad 7N_{01} = 7N_{13}$$

$$-4f_5 = -4f(\mathrm{CO_3}): \quad 4N_{02} = 4N_{10} + 4N_{11} + 4N_{12}$$

$$f_{12} + f_0 - f_3 - 7f_4 - 4f_5: \quad 0 = 0$$

For the solution (II), we have the balances:

$$f_1 = f(\mathrm{H}):$$

$$2N_1 + N_2(1+2n_2) + N_3(1+2n_3) + N_5(1+2n_5) + 2N_6n_6 + N_{10}(2+2n_{10})$$

$$+ N_{11}(1+2n_{11}) + 2N_{12}n_{12} + 2N_{25}n_{25} + N_{26}(1+2n_{26}) + 2N_{27}n_{27} = 14N_{04} + 2N_{05} + 2N_{07}$$

$$f_2 = f(\mathrm{O}):$$

$$N_1 + N_2n_2 + N_3(1+n_3) + N_5(4+n_5) + N_6(4+n_6) + N_{10}(3+n_{10}) + N_{11}(3+n_{11})$$

$$+ N_{12}(3+n_{12}) + N_{25}n_{25} + N_{26}(1+n_{26}) + N_{27}(4+n_{27}) = 11N_{04} + 4N_{05} + 2N_{06} + N_{07}$$

$$f_{12} = 2\cdot f(\mathrm{O}) - f(\mathrm{H}):$$

$$-N_2 + N_3 + 7N_5 + 8N_6 + 4N_{10} + 5N_{11} + 6N_{12} + N_{26} + 8N_{27} = 8N_{04} + 6N_{05} + 4N_{06}$$

$$f_0 = \mathrm{ChB}: \quad N_2 - N_3 - N_5 - 2N_6 - N_{11} - 2N_{12} + 2N_{25} + N_{26} = 0$$

$$-2f_3 = -2f(\mathrm{Fe}): \quad 2N_{04} = 2N_{25} + 2N_{26} + 2N_{27}$$

$$-6f_4 = -6f(\mathrm{SO_4}): \quad 6N_{04} + 6N_{05} = 6N_5 + 6N_6 + 6N_{27}$$

$$-4f_5 = -4f(\mathrm{CO_3}): \quad 4N_{06} = 4N_{10} + 4N_{11} + 4N_{12}$$

$$f_{12} + f_0 - 2f_3 - 6f_4 - 4f_5: \quad 0 = 0$$

For the solution (III), we have the balances:

$$f_1 = f(\mathrm{H}):$$
$$2N_1 + N_2(1+2n_2) + N_3(1+2n_3) + N_5(1+2n_5) + 2N_6n_6 + N_7(2+2n_7) + N_8(1+2n_8)$$
$$+ 2N_9n_9 + N_{10}(2+2n_{10}) + N_{11}(1+2n_{11}) + 2N_{12}n_{12} = 6N_{08} + 2N_{09} + 2N_{011}$$

$$f_2 = f(\mathrm{O}):$$
$$N_1 + N_2n_2 + N_3(1+n_3) + N_5(4+n_5) + N_6(4+n_6) + N_7(4+n_7) + N_8(4+n_8)$$
$$+ N_9(4+n_9) + N_{10}(3+n_{10}) + N_{11}(3+n_{11}) + N_{12}(3+n_{12}) = 6N_{08} + 4N_{09} + 2N_{010} + N_{011}$$

$$f_{12} = 2 \cdot f(\mathrm{O}) - f(\mathrm{H}):$$
$$-N_2 + N_3 + 7N_5 + 8N_6 + 6N_7 + 7N_8 + 8N_9 + 4N_{10} + 5N_{11} + 6N_{12} = 6N_{08} + 6N_{09} + 4N_{010}$$

$$f_0 = \mathrm{ChB}: \quad N_2 - N_3 - N_5 - 2N_6 - N_8 - 2N_9 - N_{11} - 2N_{12} = 0$$
$$-6f_3 = -6f(\mathrm{SO_4}): \quad 6N_{09} = 6N_5 + 6N_6$$
$$-4f_4 = -4f(\mathrm{CO_3}): \quad 4N_{010} = 4N_{10} + 4N_{11} + 4N_{12}$$
$$-6f_5 = -6f(\mathrm{C_2O_4}): \quad 6N_{08} = 6N_7 + 6N_8 + 6N_9$$

$$f_{12} + f_0 - 6f_3 - 4f_4 - 6f_5: \quad 0 = 0$$

For the solution (IV), we have the balances:

$$f_1 = f(\mathrm{H}):$$
$$2N_1 + N_2(1+2n_2) + N_3(1+2n_3) + N_5(1+2n_5) + 2N_6n_6 + N_7(2+2n_7) + N_8(1+2n_8)$$
$$+ 2N_9n_9 + N_{10}(2+2n_{10}) + N_{11}(1+2n_{11}) + 2N_{12}n_{12} + 2N_{25}n_{25} + N_{26}(1+2n_{26}) + 2N_{27}n_{27}$$
$$+ 2N_{28}n_{28} + 2N_{29}n_{29} + 2N_{39}n_{39} = 14N_{012} + 6N_{013} + 2N_{014} + 2N_{016}$$

$$f_2 = f(\mathrm{O}):$$
$$N_1 + N_2n_2 + N_3(1+n_3) + N_5(4+n_5) + N_6(4+n_6) + N_7(4+n_7) + N_8(4+n_8) + N_9(4+n_9)$$
$$+ N_{10}(3+n_{10}) + N_{11}(3+n_{11}) + N_{12}(3+n_{12}) + N_{25}n_{25} + N_{26}(1+n_{26}) + N_{27}(4+n_{27})$$
$$+ N_{28}(8+n_{28}) + N_{29}(12+n_{29}) + N_{39}(4+n_{39}) = 11N_{012} + 6N_{013} + 4N_{014} + 2N_{015} + N_{016}$$

$$f_{12} = 2 \cdot f(\mathrm{O}) - f(\mathrm{H}):$$
$$-N_2 + N_3 + 7N_5 + 8N_6 + 6N_7 + 7N_8 + 8N_9 + 4N_{10} + 5N_{11} + 6N_{12} + N_{26} + 8N_{27} + 16N_{28}$$
$$+24N_{29} + 8N_{39} = 8N_{012} + 6N_{013} + 6N_{014} + 4N_{015}$$

$$f_0 = \mathrm{ChB}:$$
$$N_2 - N_3 - N_5 - 2N_6 - N_8 - 2N_9 - N_{11} - 2N_{12} + 2N_{26} + N_{27} - 2N_{29} - 4N_{30} = 0$$
$$-6f_3 = -6f(\mathrm{SO_4}): \quad 6N_{012} + 6N_{014} = 6N_5 + 6N_6 + 6N_{28}$$
$$-4f_4 = -4f(\mathrm{CO_3}): \quad 4N_{015} = 4N_{10} + 4N_{11} + 4N_{12}$$
$$-6f_5 = -6f(\mathrm{C_2O_4}): \quad 6N_{013} = 6N_7 + 6N_8 + 6N_9 + 12N_{28} + 18N_{29} + 6N_{39}$$
$$-2f_6 = -2f(\mathrm{Fe}): \quad 2N_{012} = 2N_{25} + 2N_{26} + 2N_{27} + 2N_{28} + 2N_{29} + 2N_{39}$$

$$f_{12} + f_0 - 6f_3 - 4f_4 - 6f_5 - 2f_6: \quad 0 = 0$$

Summarizing, for all the solutions (I)–(IV), we obtain the identities 0 = 0:

$$\text{(I)}f_{12} + f_0 - f_3 - 7f_4 - 4f_5 = 0;\ \text{(II)}f_{12} + f_0 - 2f_3 - 6f_4 - 4f_5 = 0;\ \text{(III)}f_{12} + f_0 - 6f_3 - 4f_4 - 6f_5 = 0;$$
$$\text{(IV)}f_{12} + f_0 - 6f_3 - 4f_4 - 6f_5 - 2f_6 = 0 \tag{4}$$

All the solutions are non-redox systems. Except protonation/hydrolytic effects in (I)–(IV), the complexation and precipitation occur in (II) and (IV); the precipitation of $\mathbf{FeC_2O_4}$ does not occur there at sufficiently high concentrations of H_2SO_4.

The solutions can be mixed according to titrimetric mode. In particular, we refer to the D + T systems obtained in the titrations T ⇒ D indicated above, namely (I) ⇒ (II), (I) ⇒ (III) and (I) ⇒ (IV). According to the notation applied elsewhere, for example, in Refs. [19–23], V_0 mL of D is titrated with volume V mL of T, added up to a given point of the titration, and the D + T mixture with volume V_0 + V mL is formed at this point if the assumption of the volume additivity is valid.

We assume $V_1 = V$, $CV = 10^3 \cdot N_{01}/N_A$ (N_A: Avogadro's number) and $V_2 = V_0$ and $C_0V_0 = 10^3 \cdot N_{04}/N_A$ for (I) ⇒ (II); $V_3 = V_0$ and $C_0V_0 = 10^3 \cdot N_{08}/N_A$ for (I) ⇒ (III); $V_4 = V_0$ and $C_{01}V_0 = 10^3 \cdot N_{013}/N_A$ and $C_{02}V_0 = 10^3 \cdot N_{012}/N_A$ for (I) ⇒ (IV). Concentrations of the species $X_i^{z_i} \cdot n_iH_2O$ in the related systems are defined by relation $[X_i^{z_i}](V_0 + V) = N_i/N_A$, where $[X_i^{z_i}]$ is the molar concentration of $X_i^{z_i} \cdot n_iH_2O$ for i ≥ 2. The progress of the titration in (I) ⇒ (II) and (I) ⇒ (III) can be defined by the fraction titrated [24–29] value

$$\Phi = \frac{C \cdot V}{C_0 \cdot V_0} \tag{5}$$

whereas V will be taken as a parameter varied on abscissa in the graphical presentation of the system (I) ⇒ (IV).

4. Formulation of dynamic redox systems

The D and T, formed out of particular components, are considered as subsystems of the D + T system thus obtained. The titration is considered as a quasistatic process carried out under isothermal conditions and perceived both from physicochemical and analytical viewpoints.

Let us consider four starting solutions composed of:

- N_{01} molecules of $KMnO_4$, N_{02} molecules of CO_2 and N_{03} molecules of H_2O in V_1 mL of the resulting solution
- N_{04} molecules of $FeSO_4 \cdot 7H_2O$, N_{05} molecules of H_2SO_4, N_{06} molecules of CO_2 and N_{07} molecules of H_2O in V_2 mL of the resulting solution
- N_{08} molecules of $H_2C_2O_4 \cdot 2H_2O$, N_{09} molecules of H_2SO_4, N_{010} molecules of CO_2 and N_{011} molecules of H_2O in V_3 mL of the resulting solution

- N_{012} molecules of $FeSO_4 \cdot 7H_2O$, N_{013} molecules of $H_2C_2O_4{}^{\circ}2H_2O$, N_{014} molecules of H_2SO_4, N_{015} molecules of CO_2 and N_{016} molecules of H_2O in V_4 mL of the resulting solution

We start our considerations from the most complex dynamic system (I) ⇒ (IV), where V mL $KMnO_4$ (C) + CO_2 (C_1) is added into V_0 mL $FeSO_4$ (C_{01}) + $H_2C_2O_4$ (C_{02}) + H_2SO_4 (C_{03}) + CO_2 (C_{04}) at the defined point of the titration. The less complex dynamic systems (I) ⇒ (II) and (I) ⇒ (III) will be considered later as a particular case of the system (I) ⇒ (IV).

4.1. Formulation of GEB for the system (I) ⇒ (IV)

Referring to the set of species in Eq. (2), we apply $a_1 = 1$ if pr1 = $\mathbf{FeC_2O_4}$ is the equilibrium solid phase (precipitate) in the system, and $a_2 = 1$ if pr2 = $\mathbf{MnC_2O_4}$ is the equilibrium solid phase in the system; otherwise, we have $a_1 = 0$ and/or $a_2 = 0$. The elemental/core balances and ChB, formulated on the basis of the set of the species Eq. (2), are as follows:

$$
\begin{aligned}
& f_1 = f(\mathrm{H}): \\
& 2N_1 + N_2(1+2n_2) + N_3(1+2n_3) + 2N_4n_4 + N_5(1+2n_5) + 2N_6n_6 + N_7(2+2n_7) \\
& + N_8(1+2n_8) + 2N_9n_9 + N_{10}(2+2n_{10}) + N_{11}(1+2n_{11}) + 2N_{12}n_{12} + 2N_{13}n_{13} \\
& + 2N_{14}n_{14} + 2N_{15}n_{15} + N_{16}(1+2n_{16}) + 2N_{17}n_{17} + 2N_{18}n_{18} + 2N_{19}n_{19} + 2N_{20}n_{20} \\
& + N_{21}(1+2n_{21}) + 2N_{22}n_{22} + 2N_{23}n_{23} + 2N_{24}n_{24} + 2N_{25}n_{25} + N_{26}(1+2n_{26}) + 2N_{27}n_{27} \\
& + 2N_{28}n_{28} + 2N_{29}n_{29} + 2N_{30}n_{30} + N_{31}(1+2n_{31}) + N_{32}(2+2n_{32}) + N_{33}(2+2n_{33}) \\
& + 2N_{34}n_{34} + 2N_{35}n_{35} + 2N_{36}n_{36} + 2N_{37}n_{37} + 2N_{38}n_{38} + 2a_1N_{39}n_{39} + 2a_2N_{40}n_{40} \\
& = 2N_{03} + 14N_{012} + 6N_{013} + 2N_{014} + 2N_{016}
\end{aligned}
$$

$$
\begin{aligned}
& f_2 = f(\mathrm{O}): \\
& N_1 + N_2n_2 + N_3(1+n_3) + N_4n_4 + N_5(4+n_5) + N_6(4+n_6) + N_7(4+n_7) + N_8(4+n_8) \\
& + N_9(4+n_9) + N_{10}(3+n_{10}) + N_{11}(3+n_{11}) + N_{12}(3+n_{12}) + N_{13}(4+n_{13}) + N_{14}(4+n_{14}) \\
& + N_{15}n_{15} + N_{16}(1+n_{16}) + N_{17}(4+n_{17}) + N_{18}(8+n_{18}) + N_{19}(12+n_{19}) + N_{20}n_{20} \\
& + N_{21}(1+n_{21}) + N_{22}(4+n_{22}) + N_{23}(4+n_{23}) + N_{24}(8+n_{24}) + N_{25}n_{25} + N_{26}(1+n_{26}) \\
& + N_{27}(4+n_{27}) + N_{28}(8+n_{28}) + N_{29}(12+n_{29}) + N_{30}n_{30} + N_{31}(1+n_{31}) + N_{32}(2+n_{32}) \\
& + N_{33}(2+n_{33}) + N_{34}(4+n_{34}) + N_{35}(8+n_{35}) + N_{36}(4+n_{36}) + N_{37}(8+n_{37}) \\
& + N_{38}(12+n_{38}) + a_1N_{39}(4+n_{39}) + a_2N_{40}(4+n_{40}) = 4N_{01} + 2N_{02} + N_{03} \\
& + 11N_{012} + 6N_{013} + 4N_{014} + 2N_{015} + N_{016}
\end{aligned}
$$

$$
\begin{aligned}
& f_{12} = 2 \cdot f(\mathrm{O}) - f(\mathrm{H}) \\
& -N_2 + N_3 + 7N_5 + 8N_6 + 6N_7 + 7N_8 + 8N_9 + 4N_{10} + 5N_{11} + 6N_{12} + 8N_{13} + 8N_{14} + N_{16} \\
& + 8N_{17} + 16N_{18} + 24N_{19}N_{21} + 8N_{22} + 8N_{23} + 16N_{24} + N_{26} + 8N_{27} + 16N_{28} + 24N_{29} + N_{31} \\
& + 2N_{32} + 2N_{33} + 8N_{34} + 16N_{35} + 8N_{36} + 16N_{37} + 24N_{38} + 8a_1N_{39} + 8a_2N_{40} = 8N_{01} + 4N_{02} \\
& + 8N_{012} + 6N_{013} + 6N_{014} + 4N_{015}
\end{aligned}
\tag{6}
$$

http://dx.doi.org/10.5772/intechopen.69249

$$f_0 = \mathrm{ChB}$$
$$N_2–N_3 + N_4–N_5–2N_6–N_8–2N_9–N_{11}–2N_{12}–N_{13}–2N_{14} + 3N_{15} + 2N_{16} + N_{17}–N_{18} - 3N_{19} + 2N_{20} + N_{21}–2N_{24} + 2N_{25} + N_{26}–2N_{28}–4N_{29} + 3N_{30} + 2N_{31} + N_{32} + 4N_{33} + N_{34} - N_{35} + N_{36}–N_{37}–3N_{38} = 0 \quad (7)$$

$$-f_3 = -f(K): \quad N_{01} = N_4 \quad (8)$$

$$-6f_4 = -6f(\mathrm{S}) = -6f(\mathrm{SO_4}): 6N_{012} + 6N_{014} = 6N_5 + 6N_6 + 6N_{22} + 6N_{27} + 6N_{34} + 12N_{35} \quad (9)$$

$$-4f_5 = -4f(\mathrm{C}):$$
$$4N_{02} + 8N_{013} + 4N_{015} = 8N_7 + 8N_8 + 8N_9 + 4N_{10} + 4N_{11} + 4N_{12} + 8N_{17} + 16N_{18} + 24N_{19} + 8N_{23} + 16N_{24} + 16N_{28} + 24N_{29} + 8N_{36} + 16N_{37} + 24N_{38} + 8a_1N_{39} + 8a_2N_{40} \quad (10)$$

$$-3f_6 = -3f(\mathrm{Fe}):$$
$$3N_{012} = 3N_{25} + 3N_{26} + 3N_{27} + 3N_{28} + 3N_{29} + 3N_{30} + 3N_{31} + 3N_{32} + 6N_{33} + 3N_{34} + 3N_{35} + 3N_{36} + 3N_{37} + 3N_{38} + 3a_1N_{39} \quad (11)$$

$$-2f_7 = -2f(\mathrm{Mn}):$$
$$2N_{01} = 2N_{13} + 2N_{14} + 2N_{15} + 2N_{16} + 2N_{17} + 2N_{18} + 2N_{19} + 2N_{20} + 2N_{21} + 2N_{22} + 2N_{23} + 2N_{24} + 2a_2N_{40} \quad (12)$$

$$f_{12} + f_0–f_3–6f_4–4f_5–3f_6 - 2f_7:$$
$$5N_{13} + 4N_{14} + N_{15} + N_{16} + N_{012} + 2N_{013} = 5N_{01} + 2N_7 + 2N_8 + 2N_9 + N_{17} + 3N_{18} + 5N_{19} + 2N_{23} + 4N_{24} + N_{25} + N_{26} + N_{27} + 5N_{28} + 7N_{29} + 2N_{36} + 4N_{37} + 6N_{38} + 3a_1N_{39} + 2a_2N_{40} \quad (13)$$

$$5[\mathrm{MnO_4^{-1}}] + 4[\mathrm{MnO_4^{-2}}] + [\mathrm{Mn^{+3}}] + [\mathrm{MnOH^{+2}}] - (2([\mathrm{H_2C_2O_4^{-1}}] + [\mathrm{HC_2O_4^{-1}}] + [\mathrm{C_2O_4^{-2}}]) + [\mathrm{MnC_2O_4^{+1}}] + 3[\mathrm{Mn(C_2O_4)_2^{-1}}] + 5[\mathrm{Mn(C_2O_4)_3^{-3}}] + 2[\mathrm{MnC_2O_4}] + 4[\mathrm{Mn(C_2O_4)_2^{-2}}] + [\mathrm{Fe^{+2}}] + [\mathrm{FeOH^{+1}}] + [\mathrm{FeSO_4}] + 5[\mathrm{Fe(C_2O_4)_2^{-2}}] + 7[\mathrm{Fe(C_2O_4)_3^{-4}}] + 2[\mathrm{FeC_2O_4^{+1}}] + 4[\mathrm{Fe(C_2O_4)_2^{-1}}] + 6[\mathrm{Fe(C_2O_4)_3^{-3}}] + 3a_1[\mathbf{FeC_2O_4}] + 2a_2[\mathbf{MnC_2O_4}]) = 5CV/(V_0 + V) - (C_{01} + 2C_{02})V_0/(V_0 + V) \Rightarrow$$
$$2([\mathrm{H_2C_2O_4}] + [\mathrm{HC_2O_4^{-1}}] + [\mathrm{C_2O_4^{-2}}]) + [\mathrm{MnC_2O_4^{+1}}] + 3[\mathrm{Mn(C_2O_4)_2^{-1}}] + 5[\mathrm{Mn(C_2O_4)_3^{-3}}] + 2[\mathrm{MnC_2O_4}] + 4[\mathrm{Mn(C_2O_4)_2^{-2}}] + [\mathrm{Fe^{+2}}] + [\mathrm{FeOH^{+1}}] + [\mathrm{FeSO_4}] + 5[\mathrm{Fe(C_2O_4)_2^{-2}}] + 7[\mathrm{Fe(C_2O_4)_3^{-4}}] + 2[\mathrm{FeC_2O_4^{+1}}] + 4[\mathrm{Fe(C_2O_4)_2^{-1}}] + 6[\mathrm{Fe(C_2O_4)_3^{-3}}] + 3a_1[\mathbf{FeC_2O_4}] + 2a_2[\mathbf{MnC_2O_4}]–(5[\mathrm{MnO_4^{-1}}] + 4[\mathrm{MnO_4^{-2}}] + [\mathrm{Mn^{+3}}] + [\mathrm{MnOH^{+2}}]) = ((C_{01} + 2C_{02})V_0–5CV)/(V_0 + V) \quad (14)$$

Eq. (14) is the shortest/simplest form of GEB for the related system; it is, of course, different from the identity 0 = 0. On the basis of Eq. (14), one can also formulate the GEB for the system (I) ⇒ (II):

$$(KMnO_4(C) + CO_2(C_1), V) \Rightarrow (FeSO_4(C_{01}) + H_2SO_4(C_{03}) + CO_2(C_{04}), V_0) \quad (15)$$

and for the system (I) ⇒ (III):

$$(KMnO_4(C) + CO_2(C_1), V) \Rightarrow (H_2C_2O_4(C_{02}) + H_2SO_4(C_{03}) + CO_2(C_{04}), V_0) \quad (16)$$

Assuming $C_{02} = 0$, after omission of the related species involved with oxalates, from Eq. (14), we have the GEB valid for the system (I) ⇒ (II):

$$[Fe^{+2}] + [FeOH^{+1}] + [FeSO_4] - (5[MnO_4^{-1}] + 4[MnO_4^{-2}] + [Mn^{+3}] + [MnOH^{+2}]) = (C_{01}V_0 - 5CV)(V_0 + V) \quad (17)$$

Assuming $C_{01} = 0$, after omission of the related Fe-species, from Eq. (14), we have the GEB valid for the system (I) ⇒ (III)

$$2([H_2 0_2 O_4] + [HC_2O_4^{-1}] + [C_2O_4^{-2}]) + [MnC_2O_4^{+1}] + 3[Mn(C_2O_4)_2^{-1}] + 5[Mn(C_2O_4)_3^{-3}] + 2[MnC_2O_4] + 4[Mn(C_2O_4)_2^{-2}] + 2a_2[\mathbf{MnC_2O_4}] - (5[MnO_4^{-1}] + 4[MnO_4^{-2}] + [Mn^{+3}] + [MnOH^{+2}]) = (2C_{02}V_0 - 5CV)(V_0 + V) \quad (18)$$

On the other hand, Eqs. (17) and (18) are the simplest/shortest linear combinations for the related subsystems of the system (I) ⇒ (IV); both are also different from the identity, of course. For comparison, the linear combination $f_{12} + f_0 - f_3 - 6f_4 - 4f_5 - 2f_6 - 2f_7 = 0$, that is where $-2f_6$ is put for $-3f_6$, gives a more extended equation, where more components are involved. Anyway, we get here the equation, not the identity 0 = 0. It must be stressed that none of the linear combinations of these equations gives the identity. This is the general property of all redox systems, of any degree of complexity.

5. Confirmation of linear dependency of balances for non-redox systems

It can be stated that $2f(O)-f(H)$ is a linear combination of charge and elemental/core balances for non-redox systems of any degree of complexity; this means that $2f(O)-f(H)$ is not a new, independent balance in non-redox systems. From Eq. (4), we see that for non-redox systems, f_{12} can be expressed as the linear combination of other balances of the system considered:

$$(I) f_{12} = f_3 + 7f_4 + 4f_5 - f_0; \quad (II) f_{12} = 2f_3 + 6f_4 + 4f_5 - f_0; \quad (III) f_{12} = 6f_3 + 4f_4 + 6f_5 - f_0; \quad (IV) f_{12} = 6f_3 + 4f_4 + 6f_5 + 2f_6 - f_0 \quad (19)$$

Eq. (4) can be also rewritten into equivalent forms:

$$\begin{aligned}
&(\mathrm{I}) \ldots (+1)f_1 + (-2)\cdot f_2 + (+1)f_3 + (+7)f_4 + (+4)\cdot f_5 - f_0 = 0;\\
&(\mathrm{II})(+1)\cdot f_1 + (-2)\cdot f_2 + (+2)\cdot f_3 + (+6)\cdot f_4 + (+4)\cdot f_5 - f_0 = 0;\\
&(\mathrm{III})(+1)\cdot f_1 + (-2)\cdot f_2 + 2(+3)\cdot f_3 + (+4)\cdot f_4 + (+6)\cdot f_5 - f_0 = 0;\\
&(\mathrm{IV})(+1)\cdot f_1 + (-2)\cdot f_2 + (+2)\cdot f_6 + 2(+3)\cdot f_3 + (+4)\cdot f_4 + (+6)\cdot f_5 - f_0 = 0
\end{aligned} \tag{20}$$

As we see, the coefficient at the corresponding elemental/core balance in the related sum is equal to the oxidation number of the corresponding element. The linear dependence will be thus ascertained by multiplying the elemental/core balances by the appropriate oxidation numbers. After consecutive addition of the resulting balances to the sum of 2*f*(O)–*f*(H) and charge balance, followed by simplifications, the resulting sum is reduced to the identity 0 = 0. It is the simplest way of checking the linear dependency of the equations related to non-redox systems.

For redox systems, the appropriate linear combination of 2*f*(O)–*f*(H) with charge balance and elemental/core balances related to electron-non-active elements in the system in question leads to the simplest form of GEB named as generalized electron balance (GEB). It means that the GEB is a new balance, complementary/compatible with other (charge and elemental/core) balances related to the system in question.

6. Confirmation of linear independency of balances for redox systems

Applying a similar procedure, one can also state that 2*f*(O)–*f*(H) is not a linear combination of charge and elemental/core balances for redox systems of any degree of complexity; it means that 2*f*(O)–*f*(H) is a new/independent balance in redox systems.

The independency/dependency property of the balance 2*f*(O)–*f*(H) is the basis for the division of electrolytic systems into redox and non-redox systems [8, 9]. This rule is illustrated by the following examples, related to static and dynamic systems.

7. Confirmation of equivalency of approaches I and II to GEB for the system (I) ⇒ (IV)

We apply now the linear combination (algebraic sum) of Eqs. (6–9) for ChB and elemental/core balances, involving electron-non-active elements: H, O, K and S, perceived in terms of the Approach I to GEB as 'fans', we have:

$$
\begin{aligned}
& f_{12}+f_0-f_3-6f_4: \\
& 6N_7+6N_8+6N_9+4N_{10}+4N_{11}+4N_{12}+7N_{13}+6N_{14}+3N_{15}+3N_{16}+9N_{17}+15N_{18} \\
& +21N_{19}+2N_{20}+2N_{21}+2N_{22}+8N_{23}+14N_{24}+2N_{25}+2N_{26}+2N_{27}+14N_{28}+20N_{29} \\
& +3N_{30}+3N_{31}+3N_{32}+6N_{33}+3N_{34}+3N_{35}+9N_{36}+15N_{37}+21N_{38}+8a_1N_{39}+8a_2N_{40} \\
& =7N_{01}+6N_{013}+4(N_{02}+N_{015}) \Rightarrow 6(N_7+N_8+N_9)+4(N_{10}+N_{11}+N_{12})+7N_{13}+6N_{14} \\
& +3N_{15}+3N_{16}+9N_{17}+15N_{18}+21N_{19}+2(N_{20}+N_{21}+N_{22})+8N_{23}+14N_{24} \\
& +2(N_{25}+N_{26}+N_{27})+14N_{28}+20N_{29}+3(N30+N31+N32+2N33+N_{34}+N_{35}) \\
& +9N_{36}+15N_{37}+21N_{38}+8a_1N_{39}+8a_2N_{40}=7N_{01}+6N_{013}+4(N_{02}+N_{015})
\end{aligned} \tag{21}
$$

Denoting by Z_C (= 6), Z_{Fe} (= 26) and Z_{Mn} (= 25), the atomic numbers for electron-active elements ('players') C, Fe and Mn, from Eqs. (10)–(12) and (21), we have, by turns,

$$
\begin{aligned}
& Z_C \cdot f_5+Z_{Fe} \cdot f_6+Z_{Mn} \cdot f_7-(f_{12}+f_0-f_3-6f_4) \\
& (2Z_C-6)(N_7+N_8+N_9)+(Z_C-4)(N_{10}+N_{11}+N_{12})+(Z_{Mn}-7)N_{13}+(Z_{Mn}-6)N_{14} \\
& +(Z_{Mn}-3)(N_{15}+N_{16})+N_{17}(Z_{Mn}+2Z_C-9)+N_{18}(Z_{Mn}+4Z_C-15)+N_{19}(Z_{Mn}+6Z_C-21) \\
& +(Z_{Mn}-2)(N_{20}+N_{21}+N_{22})+N_{23}(Z_{Mn}+2Z_C-8)+N_{24}(Z_{Mn}+4Z_C-14)+(Z_{Fe}-2)(N_{25} \\
& +N_{26}+N_{27})(Z_{Fe}+4Z_C-12)N_{28}+(Z_{Fe}+6Z_C-20)N_{29}+(Z_{Fe}-3)(N_{30}+N_{31}+N_{32}) \\
& +2(Z_{Fe}-3)N_{33}+(Z_{Fe}-3)(N_{34}+N_{35})(Z_{Fe}+2Z_C-9)N_{36}+(Z_{Fe}+4Z_C-15)N_{37} \\
& +(Z_{Fe}+6Z_C-21)N_{38}+a_1(Z_{Fe}+2Z_C-8)N_{39}+a_2(Z_{Mn}+2Z_C-8)N_{40}=(Z_{Mn}-7)N_{01} \\
& +(Z_{Fe}-2)N_{012}+2(Z_C-3)N_{013}+(Z_C-4)(N_{02}+N_{015})
\end{aligned}
$$

$$
\begin{aligned}
& 2(Z_C-3)(N_7+N_8+N_9)+(Z_C-4)(N_{10}+N_{11}+N_{12})+(Z_{Mn}-7)N_{13}+(Z_{Mn}-6)N_{14} \\
& +(Z_{Mn}-3)(N_{15}+N_{16})+N_{17}(Z_{Mn}+2Z_C-9)+N_{18}(Z_{Mn}+4Z_C-15)+N_{19}(Z_{Mn}+6Z_C-21) \\
& +(Z_{Mn}-2)(N_{20}+N_{21}+N_{22})+N_{23}(Z_{Mn}+2Z_C-8)+N_{24}(Z_{Mn}+4Z_C-14)+(Z_{Fe}-2)(N_{25} \\
& +N_{26}+N_{27})(Z_{Fe}+4Z_C-12)N_{28}+(Z_{Fe}+6Z_C-20)N_{29}+(Z_{Fe}-3)(N_{30}+N_{31}+N_{32}) \\
& +2(Z_{Fe}-3)N_{33}+(Z_{Fe}-3)(N_{34}+N_{35})(Z_{Fe}+2Z_C-9)N_{36}+(Z_{Fe}+4Z_C-15)N_{37} \\
& +(Z_{Fe}+6Z_C-21)N_{38}+a_1(Z_{Fe}+2Z_C-8)N_{39}+a_2(Z_{Mn}+2Z_C-8)N_{40} \\
& =(Z_{Mn}-7)N_{01}+(Z_{Fe}-2)N_{012}+2(Z_C-3)N_{013}+(Z_C-4)(N_{02}+N_{015})
\end{aligned}
$$

$$
\begin{aligned}
& 2(Z_C-3)([H_2C_2O_4]+[HC_2O_4^{-1}]+[C_2O_4^{-2}])+(Z_C-4)([H_2CO_3]+[HCO_3^{-1}]+[CO_3^{-2}]) \\
& +(Z_{Fe}-2)([Fe^{+2}]+[FeOH^{+1}]+[FeSO_4])+(Z_{Fe}-3)([Fe^{+3}]+[FeOH^{+2}]+[Fe(OH)_2^{+1}] \\
& +2[Fe_2(OH)_2^{+1}]+[FeSO_4^{+1}]+[Fe(SO_4)_2^{-1}])+(Z_{Mn}-7)([MnO_4^{-1}]+(Z_{Mn}-6)[MnO_4^{-2}] \\
& +(Z_{Mn}-3)([Mn^{+3}]+[MnOH^{+2}])+(Z_{Mn}-2)([Mn^{+2}]+[MnOH^{+1}]+[MnSO_4]) \\
& +(Z_{Fe}-2+4(Z_C-3))[Fe(C_2O_4)_2^{-2}]+(Z_{Fe}-2+6(Z_C-3))[Fe(C_2O_4)_3^{-4}] \\
& +a_1(Z_{Fe}-2+2(Z_C-3))[\mathbf{FeC_2O_4}]+(Z_{Fe}-3+2(Z_C-3))[FeC_2O_4^{+1}] \\
& +(Z_{Fe}-3+4(Z_C-3))[Fe(C_2O_4)_2-1]+(Z_{Fe}-3+6(Z_C-3))[Fe(C_2O_4)_3^{-3}] \\
& +(Z_{Mn}-2+2(Z_C-3))[MnC_2O_4]+(Z_{Mn}-2+4(Z_C-3))[Mn(C_2O_4)_2^{-2}] \\
& +(Z_{Mn}-3+2(Z_C-3))[MnC_2O_4^{+1}]+(Z_{Mn}-3+4(Z_C-3))[Mn(C_2O_4)_2^{-1}] \\
& +(Z_{Mn}-3+6(Z_C-3))[Mn(C_2O_4)_3^{-3}]+a_2(Z_{Mn}-2+2(Z_C-3))[\mathbf{MnC_2O_4}] \\
& =2(Z_C-3)C_{01}V_0/(V_0+V)+(Z_{Fe}-2)C_{02}V_0/(V_0+V)+(Z_{Mn}-7)CV/(V_0+V)
\end{aligned} \tag{22}
$$

Eq. (22) is obtainable immediately according to the Approach I to GEB [19–22]. Note, for example, that:

$$N_{17}(Z_{Mn} + 2Z_C - 9) = N_{17}(Z_{Mn} - 3 + 1°2 \cdot (Z_C - 3))$$
$$N_{18}(Z_{Mn} + 4Z_C - 15) = N_{18}(Z_{Mn} - 3 + 2°2 \cdot (Z_C - 3))$$
$$N_{19}(Z_{Mn} + 6Z_C - 21) = N_{19}(Z_{Mn} - 3 + 3°2 \cdot (Z_C - 3)$$

The equation for GEB thus obtained (according to the Approach II to GEB [1, 4, 7–18, 24, 25]) is then equivalent to GEB, obtained according to the Approach I to GEB, based on the principle of the common pool of electrons introduced by electron-active elements ('players') of the system in question. For redox systems, the GEB is the inherent part of the generalized approach to electrolytic systems (GATES) [1], denoted as GATES/GEB.

8. Some generalizing remarks on GEB

The linear combination 2*f*(O)–*f*(H) of elemental balances, *f*(H) for H and *f*(O) for O, is a keystone of the overall thermodynamic knowledge on electrolytic systems. The 2*f*(O) –*f*(H) can be formulated both for non-redox and redox systems, with amphiprotic (co)solvent(s) involved. It is the basis for the Generalized Electron Balance (GEB) formulated according to the Approach II to GEB.

The principle of GEB formulation was presented for the first time in Refs. [30, 31] and then in Refs. [19–22, 31–36] as the Approach I to GEB. The GEB formulation according to the Approach I is based on the 'card game' principle, with a common pool of electrons as money, electron-active elements as players and electron-non-active elements as fans—not changing their oxidation degree, that is the fans' accounts are intact in this convention [13, 23], see an illustration below. Electrons are considered as money, transferred between players; the knowledge of oxidation numbers of all elements in the system in question is needed there.

The Approach I to GEB, named also as the 'short' version of GEB, needs a knowledge of oxidation numbers for all elements in the species participating in the system that is considered. The equivalency of the Approaches I and II means that the equation obtained by a suitable linear combination of pr-GEB with charge balance and other elemental/core balances becomes identical with the one obtained directly from the Approach I to GEB.

Although derivation of GEB according to the Approach II is more extensive/laborious, it enables to formulate this balance without prior knowledge of oxidation numbers for the elements involved in the system. It is the paramount advantage of the Approach II to GEB, particularly when applied to more complex organic species, with radicals and ion-radicals involved. Moreover, within the Approaches I and II, the roles of oxidants and reducers are not ascribed a priori to particular components forming the redox system and the species formed in this system.

Ultimately, GEB, charge and elemental/core balances are expressed in terms of molar concentrations—to be fully compatible with expressions for equilibrium constants, interrelating molar concentrations of defined species on the basis of the mass action law applied to the correctly written reaction equation. The law of mass action is the one and only chemical law applied in GATES.

GEB is perceived as the law of matter conservation, as the general law of nature related to electrolytic (aqueous, non-aqueous or mixed-solvent media) redox systems and as a synthesis of physical and chemical laws [1, 14, 15, 23, 24, 27]. This law can also be extended on the systems with mixed (e.g., binary) solvents with amphiprotic (protophilic and protogenic) and aprotic properties. GEB is a rather unexpected consequence of the concentration balances for H and O, and therefore the formulation of GEB, especially as the Approach II to GEB, is regarded as the scientific discovery and not as a confirmation of the obvious fact arising from other, fundamental laws of nature. This fact is emphasized in this chapter in the context of philosophical understanding of the scientific discoveries in the aspect of the laws of nature.

The GEB, together with charge and concentration/core balances and a set of independent equilibrium constants, provides a complete set of equations used for a thermodynamic description of a redox system taken for quantitative considerations within GATES/GEB $\in$ GATES.

The roles of oxidants and reductants are not ascribed a priori to particular components forming the redox system and to the species formed in this system. In other words, full 'democracy' is established a priori within GATES/GEB.

The Approach II to GEB shows that the equivalent equations for GEB are derived from the common root of the elements conservation and then GEB is fully compatible with charge and concentration balances like 'the lotus flower, lotus leaf and lotus seed come from the same root' [13]. This compatibility is directly visible from the viewpoint of the Approach II to GEB. The GEB, based on a reliable law of the matter conservation, is equally robust as equations for charge and concentration balances. The complementarity of the GEB (Approaches I and II) to other balances is regarded as the expression of harmony of nature, and GATES/GEB is an example of excellent epistemological paradigm [27].

The number of electron-active elements (considered as players, in terms of Approach I to GEB) in a redox system, considered according to GATES/GEB principles, is practically unlimited; among others, the systems with three [24] or four [1] players were considered.

In the modeling of real systems, it is assumed that an effect of the matter (such as H_2O, CO_2 and O_2) exchange with the environment is negligibly small within the period designed for certain chemical operations made on the system.

9. Completion of balances

The set of balances for the system (I) $\Rightarrow$ (IV) is composed of GEB (e.g., 14 or 22) and equations obtained from the balances (7)–(12) are expressed in terms of molar concentrations, namely:

$$[H^{+1}] - [OH^{-1}] + [K^{+1}] - [HSO_4^{-1}] - 2[SO_4^{-2}] - [HC_2O_4^{-1}] - 2[C_2O_4^{-2}] - [HCO_3^{-1}] - 2[CO_3^{-2}] - [MnO_4^{-1}]$$
$$-2[MnO_4^{-2}] + 3[Mn^{+3}] + 2[MnOH^{+2}] + [MnC_2O_4^{+1}] - [Mn(C_2O_4)_2^{-1}] - 3[Mn(C_2O_4)_3^{-3}] + 2[Mn^{+2}]$$
$$+ [MnOH^{+1}] - 2[Mn(C_2O_4)_2^{-2}] + 2[Fe^{+2}] + [FeOH^{+1}] - 2[Fe(C_2O_4)_2^{-2}] - 4[Fe(C_2O_4)_3^{-4}] + 3[Fe^{+3}]$$
$$+ 2[FeOH^{+2}] + [Fe(OH)_2^{+1}] + 4[Fe_2(OH)_2^{+4}] + [FeSO_4^{+1}] - [Fe(SO_4)_2^{-1}] + [FeC_2O_4^{+1}]$$
$$-[Fe(C_2O_4)_2^{-1}] - 3[Fe(C_2O_4)_3^{-3}] = 0 \quad (23)$$

$$[K^{+1}] = CV/(V_0 + V) \quad (24)$$

CB(S) :
$$[HSO_4^{-1}] + [SO_4^{-2}] + [MnSO_4] + [FeSO_4] + [FeSO_4^{+1}] + 2[Fe(SO_4)_2^{-1}] = (C_{01} + C_{03})V_0/(V_0 + V) \quad (25)$$

CB(C) :
$$2([H_2C_2O_4] + [HC_2O_4^{-1}] + [C_2O_4^{-2}]) + [H_2CO_3] + [HCO_3^{-1}] + [CO_3^{-2}]$$
$$+2[MnC_2O_4] + 4[Mn(C_2O_4)_2^{-2}] + 4[Fe(C_2O_4)2^{-2}] + 6[Fe(C_2O_4)_3^{-4}] + 2[FeC_2O_4^{+1}] \quad (26)$$
$$+4[Fe(C_2O_4)_2^{-1}] + 6[Fe(C_2O_4)_3^{-3}] + 2a_1[\mathbf{FeC_2O_4}] + 2a_2[\mathbf{MnC_2O_4}]$$
$$= ((2C_{02} + C_{04})V_0 + C_1V)/(V_0 + V)$$

CB(Fe) :
$$[Fe^{+2}] + [FeOH^{+1}] + [FeSO_4] + [Fe(C_2O_4)_2^{-2}] + [Fe(C_2O_4)_3^{-4}] + [Fe^{+3}] + [FeOH^{+2}]$$
$$+[Fe(OH)_2^{+1}] + 2[Fe_2(OH)_2^{+4}] + [FeSO_4^{+1}] + [Fe(SO_4)_2^{-1}] + [FeC_2O_4^{+1}] + [Fe(C_2O_4)2^{-1}] \quad (27)$$
$$+[Fe(C_2O_4)3^{-3}] + a_1[\mathbf{FeC_2O_4}] = C_{01}V_0/(V_0 + V)$$

CB(Mn) :
$$[MnO_4^{-1}] + [MnO_4^{-2}] + [Mn^{+3}] + [MnOH^{+2}] + [MnC_2O_4^{+1}] + [Mn(C_2O_4)2^{-1}]$$
$$+[Mn(C_2O_4)3^{-3}] + [Mn^{+2}] + [MnOH^{+1}] + [MnSO_4] + [MnC_2O_4] + [Mn(C_2O_4)2^{-2}] \quad (28)$$
$$+a_2[\mathbf{MnC_2O_4}] = CV/(V_0 + V)$$

The balances can be specified as equations or equalities. The equality is represented here by relation (24), where only one species is involved. In Eqs. (23) and (25)–(28), we have concentrations of more species, interrelated in expressions for equilibrium constants, formulated on the basis of the proper stoichiometric reaction notations. As with above results, we have seven balances: six equations and one equality for the system (I) ⇒ (IV); the equality (24) can enter immediately the charge balance (23). The current volume V of titrant T added is a parameter (steering variable) in these balances.

10. The arrangement of relations for equilibrium constants

The balances written in terms of molar concentrations are congruent with a complete set of independent (non-contradictory [9, 16]) relations for the equilibrium constants, interrelating concentrations of some species in the balances.

The complete set of equilibrium constants, namely ionic product of water, dissociation constants, standard potentials, stability constants of complexes and interrelated concentrations of the species (except K^{+1}) is presented in Eq. (2).

$$\{1\}[H^{+1}][OH^{-1}] = 10^{-14}; \{2\}[HSO_4{}^{-1}] = 10^{1.8}[H^{+1}][SO_4{}^{-2}]; \{3\}[H_2C_2O_4] = 10^{5.2}[H^{+1}][C_2O_4{}^{-2}]; \{4\}[HC_2O_4{}^{-1}] = 10^{3.8}[H^{+1}][C_2O_4{}^{-2}];$$

$$\{5\}[H_2CO_3] = 10^{16.4}[H^{+1}]^2[CO_3{}^{-2}]; \{6\}[HCO_3{}^{-1}] = 10^{10.1}[H^{+1}][CO_3{}^{-2}]; \{7\}[H_2CO_3] = [H_2C_2O_4{}^{0.5^\circ}10^{A(E+0.396)+pH};$$

$$\{8\}[MnO_4{}^{-1}] = [Mn^{+2}]\cdot 10^{5A(E-1.507)+8pH}; \{9\}[MnO_4{}^{-2}] = [Mn^{+2}]\cdot 10^{4A(E-1.743)+8pH}; \{10\}[Mn^{+3}] = [Mn^{+2}]\cdot 10^{A(1.509)};$$

$$[11][Fe^{+3}] = [Fe^{+2}]\cdot 10^{A(E-0.771)}; \{12\}[FeOH^{+1}] = 10^{4.5}[Fe^{+2}][OH^{-1}]; \{13\}[FeSO_4] = 10^{2.3}[Fe^{+2}][SO_4{}^{-2}];$$

$$\{14\}[Fe(C_2O_4)2^{-2}] = 10^{4.52}[Fe^{+2}][C_2O_4{}^{-2}]^2; \{15\}[Fe(C_2O_4)3^{-4}] = 10^{5.22}[Fe^{+2}][C_2O_4{}^{-2}]^3; \{16\}[FeOH^{+2}] = 10^{11.0}[Fe^{+3}][OH^{1}];$$

$$\{17\}[Fe({}_{OH)2}{}^{+1}] = 10^{21.7}[Fe^{+3}][{}^{OH^{-1}]2}; \{18\}[Fe_2({}_{OH)2}{}^{+4}] = 10^{25.1}[{}^{Fe^{+3}]2}[{}^{OH^{-1}]2}; \{19\}[FeSO_4{}^{+1}] = 10^{4.18}[Fe^{+3}][SO_4{}^{-2}];$$

$$\{20\}[Fe({}_{SO_4)2}{}^{-1}] = 10^{7.4}[Fe^{+3}][SO_4{}^{-2}]^2; \{21\}[FeC_2O_4{}^{+1}] = 10^{7.53}[Fe^{+3}][C_2O_4{}^{-2}]; \{22\}[Fe(C_2O_4)2^{-1}] = 10^{13.64}[Fe^{+3}][C_2O_4{}^{-2}]^2;$$

$$\{23\}[Fe(C_2O_4)3^{-3}] = 10^{18.46}[Fe^{+3}][C_2O_4{}^{-2}]^3; \{24\}[MnOH^{+1}] = 10^{3.4}[Mn^{+2}][OH^{-1}]; \{25\}[MnSO_4]$$

$$\{26\}[MnC_2O_4] = 10^{3.82}[Mn^{+2}][C_2O_4{}^{-2}]; \{27\}[Mn(C_2O_4)2^{-2}] = 10^{5.25}[Mn^{+2}][C_2O_4{}^{-2}]^2; \{28\}[MnOH^{+2}] = 10^{14.2}[Mn^{+3}][OH^{-1}];$$

$$\{29\}[MnC_2O_4{}^{+1}] = 10^{9.98}[Mn^{+3}][C_2O_4{}^{-2}]; \{30\}[Mn(C_2O_4)2^{-1}] = 10^{16.57}[Mn^{+3}][C_2O_4{}^{-2}]^2; \{31\}[Mn(C_2O_4)3^{-3}]$$

$$\{32\}[Fe^{+2}][C_2O_4{}^{-2}] = 10^{-6.7}\text{for pr1} = \mathbf{FeC_2O_4} \text{ and }\}7B33\ \}7D[Mn^{+2}][C_2O_4{}^{-2}] = 10^{-5.3}\text{for pr2} = \mathbf{MnC_2O_4} \tag{29}$$

11. Relation between the numbers of species, balances and equilibrium constants

For any electrolytic system, one can define the relationship between the numbers of (i) kinds of species (P) (with free H_2O molecules included), (ii) independent balances (Q) and (iii) independent equilibrium constant (R) values.

We refer first to non-redox systems discussed in Section 3, and we have:

- for the system I: P = 8, Q = 4 (from f_0, f_3, f_4, f_5) and R = 3 ({1,5,6})
- for the system II: P = 11, Q = 4 (from f_0, f_3, f_4, f_5) and R = 6 ({1,2,5,6,12,13})
- for the system III: P = 11, Q = 4 (from f_0, f_3, f_4, f_5) and R = 6 ({1–6})
- for the system IV: P = 17, Q = 5 (from f_0, f_3, f_4, f_5, f_6) and R = 11 ({1-–6,12–15,32})

Referring now to the redox systems, we have:

- for the system (I) ⇒ (II): P = 25, Q = 7 (GEB,ChB,*CB*(K),*CB*(CO_3),*CB*(S),*CB*(Fe),*CB*(Mn)) and R = 18 ({1,2,3,4,5,6,12,13,14,15,32})
- for the system (I) ⇒ (III): P = 25, Q = 6 (GEB,ChB,*CB*(K),*CB*(C),*CB*(S),*CB*(Mn)) and R = 19 ({1–10,24–31,33})
- for the system (I) ⇒ (IV): P = 40 (collected in 2), Q = 7 (GEB,ChB,*CB*(K),*CB*(C),*CB*(S),*CB*(Fe),*CB*(Mn)) and R = 33 (collected in 29)

On this basis, one can state the relationships:

- P = Q + R + 1—for non-redox systems;
- P = Q + R—for redox systems.

To standardize this relationship, it is (informally) assumed that the electron is one of the species in the redox systems. In this way, the number of species is increased by 1, and we can suggest the relationship

$$P = Q + R + 1 \tag{30}$$

as common for both redox and non-redox electrolytic systems, regardless of their degree of complexity. The relation (30), applicable in resolution of electrolytic systems, was first presented in Ref. [37]; it can be perceived as a useful counterpart of the Gibbs' phase rule (of a similar 'degree of complexity') in this area of the knowledge.

It should be noted that the total number, P = 40, of kinds of species involved in the system (I) ⇒ (IV) is relatively high.

12. The steps realized within GATES/GEB

Modeling the electrolytic systems according to GATES/GEB consists of several interacting steps [1]: (1) collection of preliminary data; (2) preparation of computer programs; (3) calculations and data handling and (4) knowledge gaining; all the steps are indicated in **Figure 1**.

12.1. Collection of preliminary data

The necessary physicochemical knowledge is mainly attainable in tables of physicochemical data, exemplified by monographs [38–40]. It should be noted that the period of interest in this

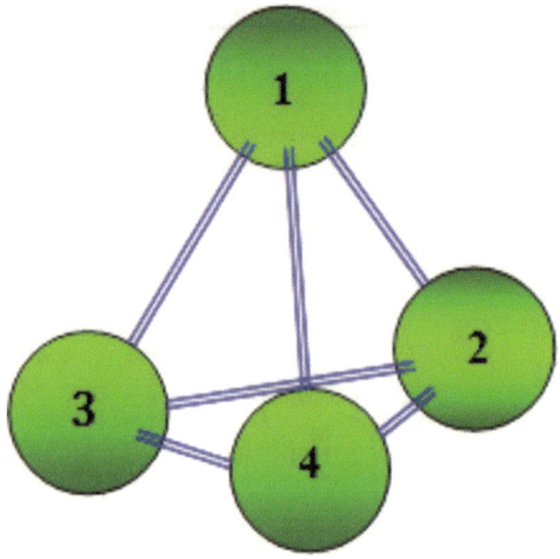

Figure 1. The steps of modeling any electrolytic system [1].

field of research is currently the past. On the other hand, the physicochemical constants originate from works released over several decades, which are clearly seen in Ref. [40], where the relevant information is included. The point is that these physicochemical constants were determined using models that had been adapted to current computing capabilities, especially in the pre-computer era; these calculations were based (exclusively, in principle) on the reaction stoichiometry. For example, the solubility products were determined on the basis of molar solubility, see for example, the remark in Refs. [41, 42], without checking whether the precipitate is the equilibrium solid phase in the system [43–48]. Acid dissociation constants were mainly determined on the basis of the Ostwald's formula (see e.g., [49, 50]). Conditional ('formal', not normal) potentials were determined for many redox pairs [51]. In the computer era, some new models resolved with use of iterative methods were elaborated. The assumptions and implementation of these models in relevant experimental studies aroused a number of concerns expressed, among others, in Refs. [52, 53]. Despite these circumstances, GATES and GATES/GEB, in particular, provides a new and reliable tool, applicable for physicochemical knowledge gaining. Thanks to this tool, it will be possible, in the immediate future, for a renaissance of interest in this—so important, after all!—field of fundamental research, which cannot be creatively developed on the basis of the previous 'paradigm' [27] based on the stoichiometry concept.

12.2. Preparation of computer program

Modeling of electrolytic systems can be realized with the use of iterative computer programs, for example MATLAB, perceived as a universal and high-level programming language [54, 55]. From the viewpoint of the form of mathematical models, MATLAB is focused on the matrix algebra procedure. MATLAB allows to make a quick and accurate computation and visualization of numerical data.

The iterative computer programs, written in MATLAB language, are exemplified in Refs. [11, 12, 16].

12.3. Calculations and data handling

The calculations can be made at different levels of the preliminary, physicochemical knowledge about the system in question. What is more, some 'variations on the subject' can also be done for this purpose; it particularly refers to metastable and non-equilibrium systems. A special emphasis will be put on complex redox systems, where all types of elementary chemical reactions proceed simultaneously and/or sequentially. In all instances, one can follow measurable quantities (potential E, pH) in dynamic and static processes and gain the information about many details not measurable in real experiments; it particularly refers to dynamic speciation.

We refer here to dynamic (titration) redox systems, represented by the system of 2 + k nonlinear equations composed of GEB, charge balance and k (≥ 1) concentration balances. The results of calculations, made with the use of an iterative computer program, are presented graphically. Thus, the plots E = E(Φ) and pH = pH(Φ) for potential E and pH of the solution and $\log[X_i^{z_i}]$ versus Φ (Eq. (5)) relationships (speciation curves) will be drawn.

The Φ concept is used for simpler systems, providing a kind of normalization (independence on V_0) in the systems considered. Φ plays a key role in the formulation of the generalized equivalence mass (GEM) concept, introduced also by Michałowski [25]. In more complex systems, the volume V is put on the abscissa.

The numerical data can be visualized in the form of two- or three-dimensional graphs (*2D*, *3D*), see for example Ref. [12].

12.4. Computer program for the system (I) ⇒ (III)

The set of independent equilibrium constants is involved in the algorithm needed for calculation purposes [14], realized in the system (I) ⇒ (IV) as specified above. An algorithm is a well-defined procedure, expressed by a sequence of unambiguous instructions, which allows a computer to solve a problem according to a computer program implemented for this purpose. The term 'unambiguous' indicates that there is no room for subjective interpretation.

The system (I) ⇒ (IV) and its subsystems (I) ⇒ (II) and (I) ⇒ (III) were simulated using an iterative computer program (MATLAB software, included in the optimization toolbox™). In particular, the computer program for the system (I) ⇒ (III) is as follows.

```
function F = Function_KMnO4_Na2C2O4(x)

global V Vmin Vstep Vmax V0 C C1 C0 C01 C02 fi H OH pH E Kw pKw A aa

global H2C2O4 HC2O4 C2O4 H2CO3 HCO3 CO3 K

global logH2C2O4 logHC2O4 logC2O4 logH2CO3 logHCO3 logCO3 logK
```

```
global Mn7O4 Mn6O4 HSO4 SO4 Na
global logMn7O4 logMn6O4 logHSO4 logSO4 logNa
global Mn3 Mn3OH Mn3C2O4 Mn3C2O42 Mn3C2O43
global logMn3 logMn3OH logMn3C2O4 logMn3C2O42 logMn3C2O43
global Mn2 Mn2OH Mn2SO4 Mn2C2O4 Mn2C2O42
global logMn2 logMn2OH logMn2SO4 logMn2C2O4 logMn2C2O42
global pr logpr q logq

pH=x(1);
E=x(2);
if aa==0
   Mn2=10.^-x(3);
   pr=0;
else
   pr=10.^-x(3);
end;
H2C2O4=10.^-x(4);
SO4=10.^-x(5);

H=10.^-pH;
pKw=14;
Kw=10.^-14;
OH=Kw./H;
A=16.92;
ZMn=25;
ZC=6;
Ksp=10.^-5.3;

HC2O4=10.^(pH-1.25).*H2C2O4;
C2O4=10.^(pH-4.27).*HC2O4;
H2CO3=10.^(A.*(E+0.386)).*H2C2O4.^0.5;

HCO3=10.^(pH-6.3).*H2CO3;
```

```
CO3=10.^(pH-10.1).*HCO3;
if aa==1
   Mn2=Ksp./C2O4;
end;
HSO4=10.^(1.8-pH).*SO4;
Mn7O4=Mn2.*10.^(5.*A.*(E-1.507)+8.*pH);
Mn6O4=10.^(A.*(0.56-E)).*Mn7O4;
Mn2OH=10.^3.4.*Mn2.*OH;
Mn2SO4=10.^2.28.*Mn2.*SO4;
Mn2C2O4=10.^3.82.*Mn2.*C2O4;
Mn2C2O42=10.^5.25.*Mn2.*C2O4.^2;
Mn3=Mn2.*10.^(A.*(E-1.509));
Mn3OH=10.^(pH-0.2).*Mn3;
Mn3C2O4=10.^9.98.*Mn3.*C2O4;
Mn3C2O42=10.^16.57.*Mn3.*C2O4.^2;
Mn3C2O43=10.^19.42.*Mn3.*C2O4.^3;
K=C.*V./(V0+V);
Na=C0.*V0./(V0+V);
   %Charge balance
F=[(H-OH+K+Na-HSO4-2.*SO4-HC2O4-2.*C2O4-HCO3-2.*CO3-Mn7O4-2.*Mn6O4...
   +3.*Mn3+2.*Mn3OH+Mn3C2O4-Mn3C2O42-3.*Mn3C2O43+2.*Mn2+Mn2OH...
   -2.*Mn2C2O42);
   %Concentration balance of Mn
   (Mn7O4+Mn6O4+Mn3+Mn3OH+Mn3C2O4+Mn3C2O42+Mn3C2O43...
   +Mn2+Mn2OH+Mn2SO4+Mn2C2O4+Mn2C2O42+aa.*pr-C.*V./(V0+V));
   %Concentration balance of C
   (2.*H2C2O4+2.*HC2O4+2.*C2O4+H2CO3+HCO3+CO3+2.*Mn2C2O4...
   +4.*Mn2C2O42+2.*Mn3C2O4+4.*Mn3C2O42+6.*Mn3C2O43...
```

```
    +2.*aa.*pr-(2.*C0.*V0+C02.*V0+C1.*V)./(V0+V));
    %Concentration balance of S
    (HSO4+SO4+Mn2SO4-C01.*V0./(V0+V));

    %Electron balance
    ((ZMn-7).*Mn7O4+(ZMn-6).*Mn6O4+(ZMn-3).*(Mn3+Mn3OH)...
    +(ZMn-2).*(Mn2+Mn2OH+Mn2SO4)+(ZC-4).*(H2CO3+HCO3+CO3)...
    +2.*(ZC-3).*(H2C2O4+HC2O4+C2O4)+(ZMn-3+2.*ZC-6).*Mn3C2O4...
    +(ZMn-3+4.*ZC-12).*Mn3C2O42+(ZMn-3+6.*ZC-18).*Mn3C2O43...
    +(ZMn-2+2.*ZC-6).*Mn2C2O4+(ZMn-2+4.*ZC-12).*Mn2C2O42+...
    +(ZMn-2+2.*ZC-6).*aa.*pr...
    -((2.*ZC-6).*C0.*V0+(ZC-4).*C02.*V0+(ZC-4).*C1.*V...
    +(ZMn-7).*C.*V)./(V0+V))];

q=Mn2.*C2O4./Ksp;

logMn2=log10(Mn2);
logMn2OH=log10(Mn2OH);
logMn2SO4=log10(Mn2SO4);
logq=log10(q);
logpr=log10(pr);
logMn2C2O4=log10(Mn2C2O4);
logMn2C2O42=log10(Mn2C2O42);
logMn3=log10(Mn3);
logMn3OH=log10(Mn3OH);
logMn3C2O4=log10(Mn3C2O4);
logMn3C2O42=log10(Mn3C2O42);
logMn3C2O43=log10(Mn3C2O43);
logMn6O4=log10(Mn6O4);
logMn7O4=log10(Mn7O4);
logH2CO3=log10(H2CO3);
```

http://dx.doi.org/10.5772/intechopen.69249

```
logHCO3=log10(HCO3);
logCO3=log10(CO3);
logH2C2O4=log10(H2C2O4);
logHC2O4=log10(HC2O4);
logC2O4=log10(C2O4);
logHSO4=log10(HSO4);
logSO4=log10(SO4);
logNa=log10(Na);
logK=log10(K);
%The end of program
```

13. Graphical presentation of the data

The results of calculations in the system (I) ⇒ (IV) are presented in **Figures 2–4**. More detailed, numerical data are specified in Ref. [14].

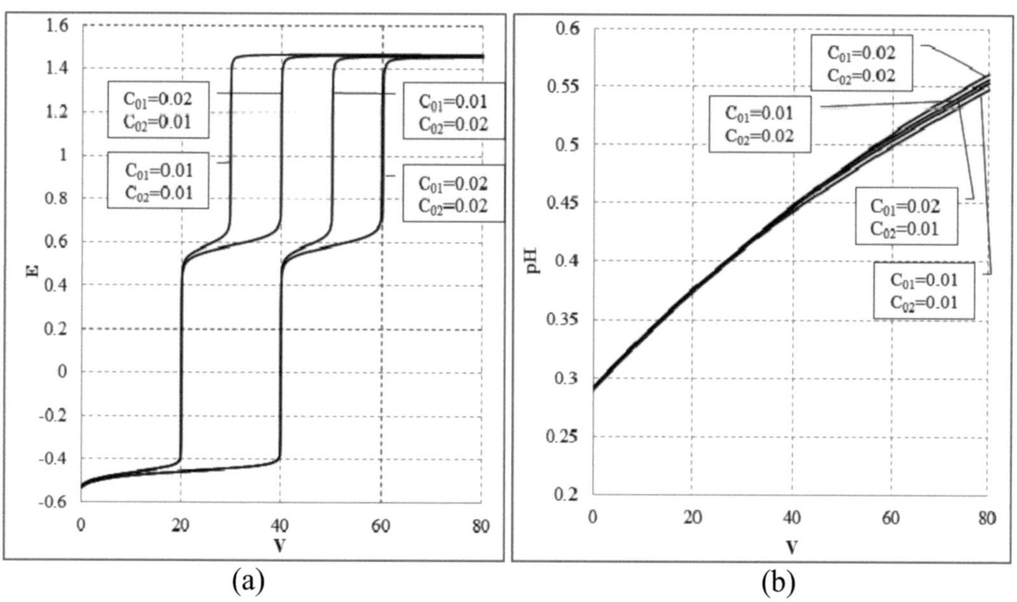

Figure 2. The (a) E versus V and (b) pH versus V relationships plotted at $V_0 = 100$, $C = 0.02$, $C_{03} = 0.5$, $C_1 = C_{04} = 0.001$ and indicated pairs of C_{01} and C_{02} values.

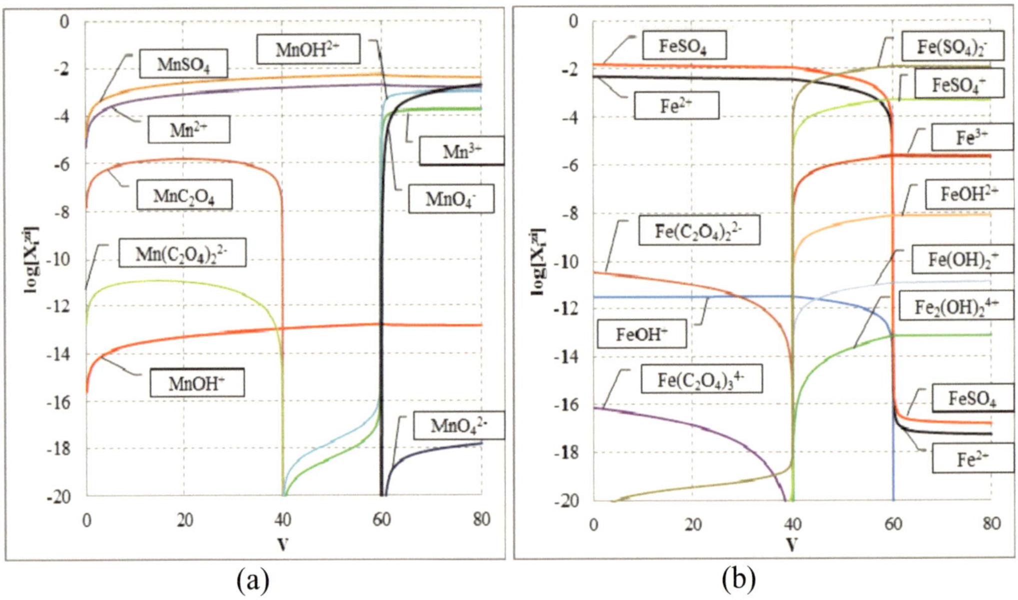

Figure 3. The speciation curves plotted for (a) Mn and (b) Fe species at $V_0 = 100$, $C = 0.02$, $C_{03} = 0.5$, $C_1 = C_{04} = 0.001$ and $C_{01} = C_{02} = 0.002$.

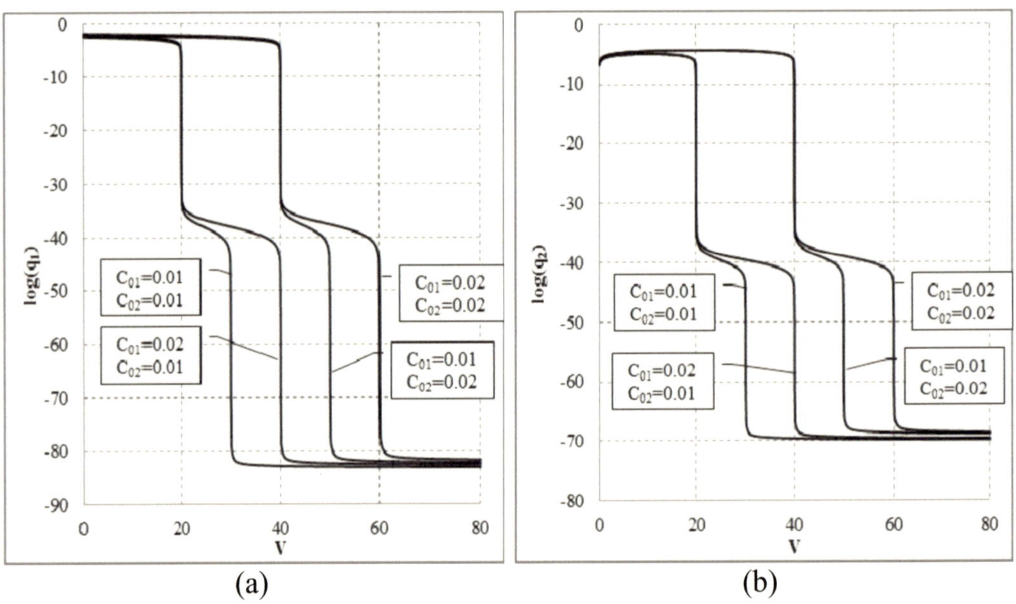

Figure 4. The $\log(q_i)$ versus V relationships (see Eq. (31)) plotted for (a) Fe (i = 1) and (b) Mn (i = 2) oxalates.

There are valid relationships: $CV_{eq1} = 0.2 \cdot C_{01}V_0$ for iron and $CV_{eq2} = 0.4 \cdot C_{02}V_0$ for oxalate. For $V_0 = 100$, $C = 0.02$ we have, in particular, $V_{eq1} = 10$ and $V_{eq2} = 20$ at $C_{01} = 0.01$, $C_{02} = 0.01$ and $V_{eq1} = 20$ and $V_{eq2} = 40$ at $C_{01} = 0.02$ and $C_{02} = 0.02$. This agrees exactly with the position of the points

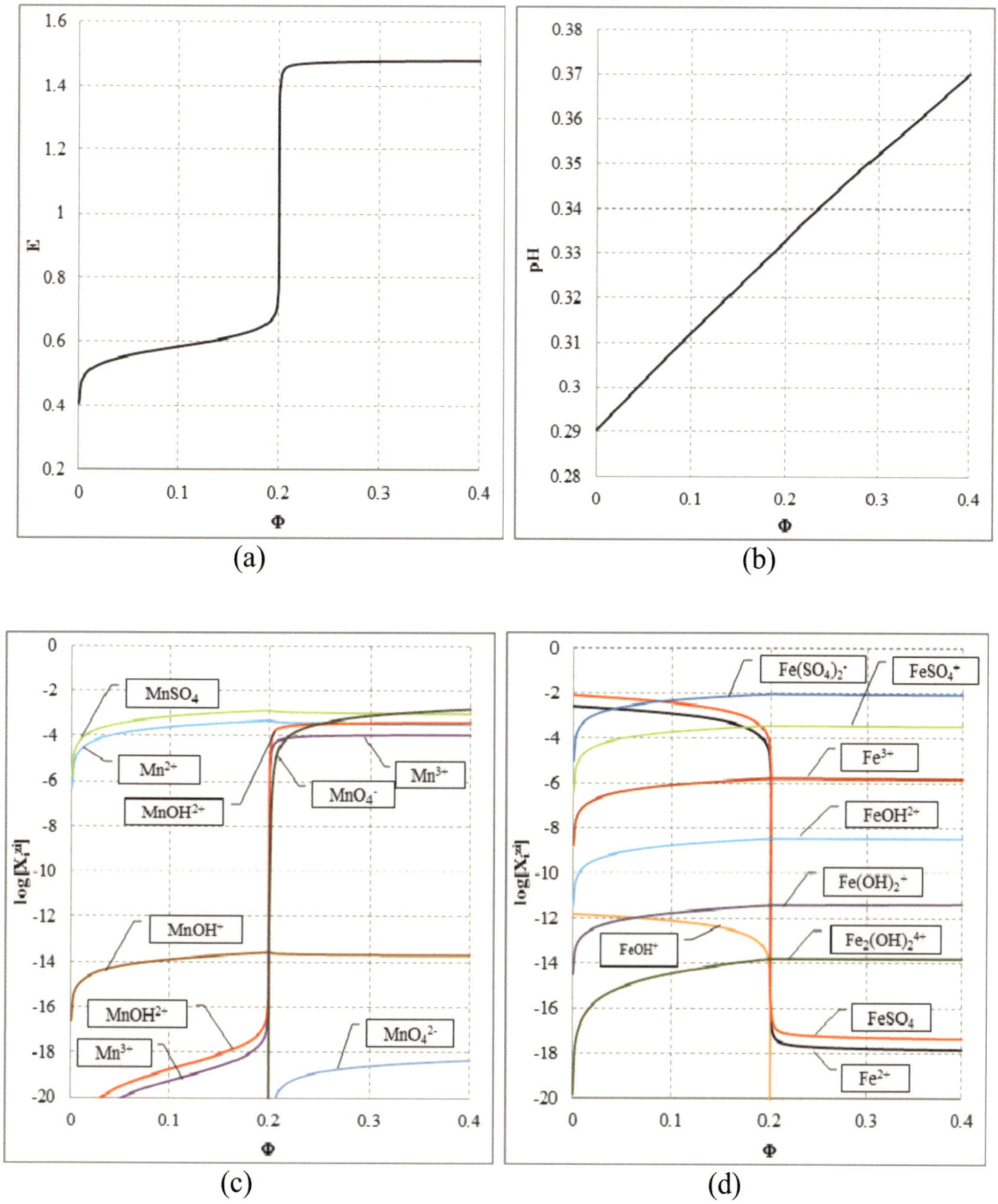

Figure 5. The relationships: (a) E = E(Φ), (b) pH = pH(Φ) and the speciation curves $\log[X_i^{z_i}]$ versus Φ for (c) manganese and (d) iron species plotted for titration of $V_0 = 100$ mL of $C_0 = 0.01$ mol/L $FeSO_4$ + C_{01} mol/L H_2SO_4 as D with V mL of C = 0.02 mol/L $KMnO_4$ as T.

where jumps of potential E occur and should be compared with the plots of titration curves for individual analytes: $FeSO_4$ (**Figure 5a**) and $H_2C_2O_4$ (**Figure 6a**), where abscissas are expressed in terms of the fraction titrated Φ (Eq. (5)). The related pH versus Φ relationships are presented in **Figures 5b** and **6b**. To explain/formulate the reactions occurred in the systems together with

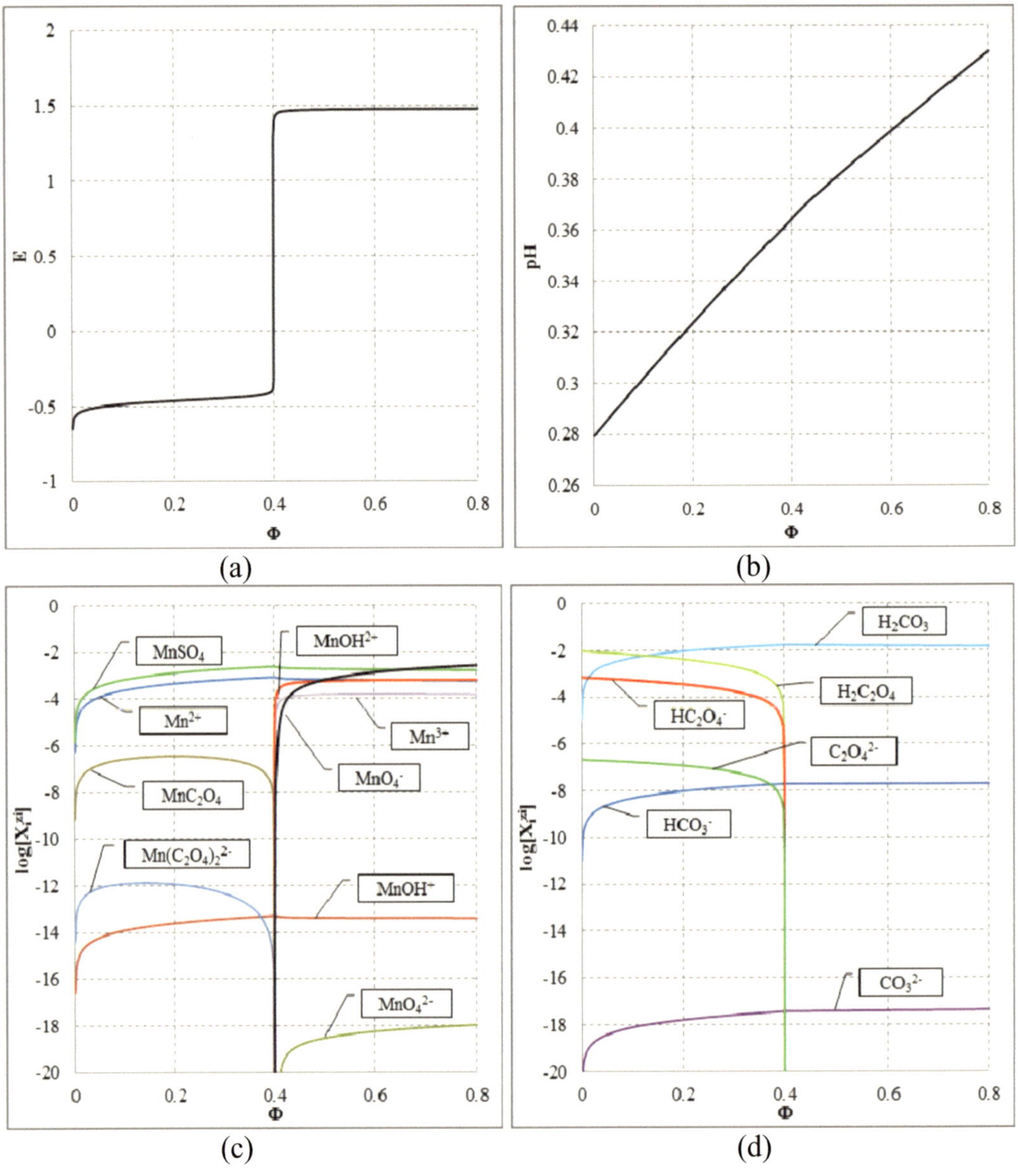

Figure 6. The relationships: (a) E = E(Φ), (b) pH = pH(Φ) and the speciation curves $\log[X_i^{z_i}]$ for (c) Mn-, (d) C-species for titration of V_0 = 100 mL of C_0 = 0.01 mol/L $H_2C_2O_4$ + C_{01} mol/L H_2SO_4 as D with V mL of C = 0.02 mol/L $KMnO_4$ as T.

their relative efficiencies, the speciation diagrams depicted in **Figures 3a**, **b**, and **5c**, **d** are used. From **Figure 4a** and **b**, we see that

$$q_1 = [Fe^{+2}][C_2O_4^{-2}]/K_{sp1} < 1 \quad \text{and} \quad q_2 = [Mn^{+2}][C_2O_4^{-2}]/K_{sp2} < 1 \tag{31}$$

i.e., the precipitates $\mathbf{FeC_2O_4}$ (K_{sp1}) and $\mathbf{MnC_2O_4}$ (K_{sp2}) do not exist in this system as the equilibrium solid phases at the pre-assumed sufficiently high concentration C_{03} of H_2SO_4; $\mathbf{MnO_2}$ is not formed there as well, i.e., $a_1 = a_2 = 0$ in Eqs. (26)–(28).

14. Deficiency and veracity of equilibrium data

In some 'variations on the subject', we try to know what would happen if some constraints put on the metastable system are removed and the reaction is conducted in a thermodynamic manner, in accordance with the conditions imposed by the equilibrium constants [1, 36]. One can also analyze the data resulting from (intentional) omission or (factual, presupposed) incomplete physicochemical knowledge on the system studied.

Some computer simulations can be used to check some effects involved with complexation phenomena. For example, we intend to check the effect involved with formation of sulfate complexes $FeSO_4$, $MnSO_4$ and (particularly) $FeSO_4^{+1}$, $Fe(SO_4)_2^{-1}$ in the system (I) ⇒ (II). The shapes of the titration curves E = E(Φ) are compared in **Figure 7**.

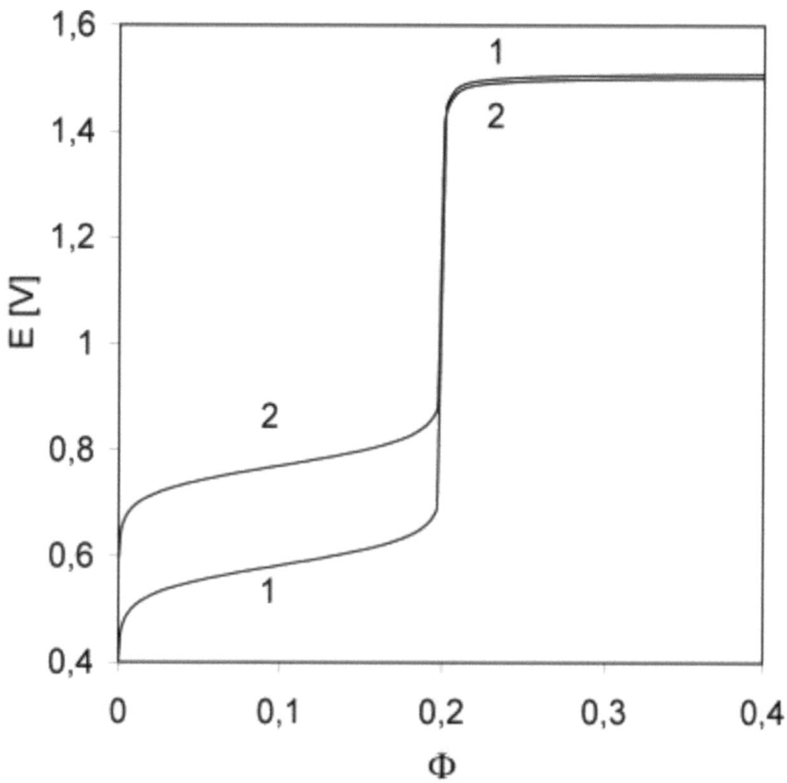

Figure 7. The E versus Φ relationships plotted for the system (I) ⇒ (II): (1) at pre-assumed physicochemical knowledge and (2) after intentional omission of all sulfate complexes; $C_0 = 0.01$, $C_{01} = 1.0$ and $C = 0.02$.

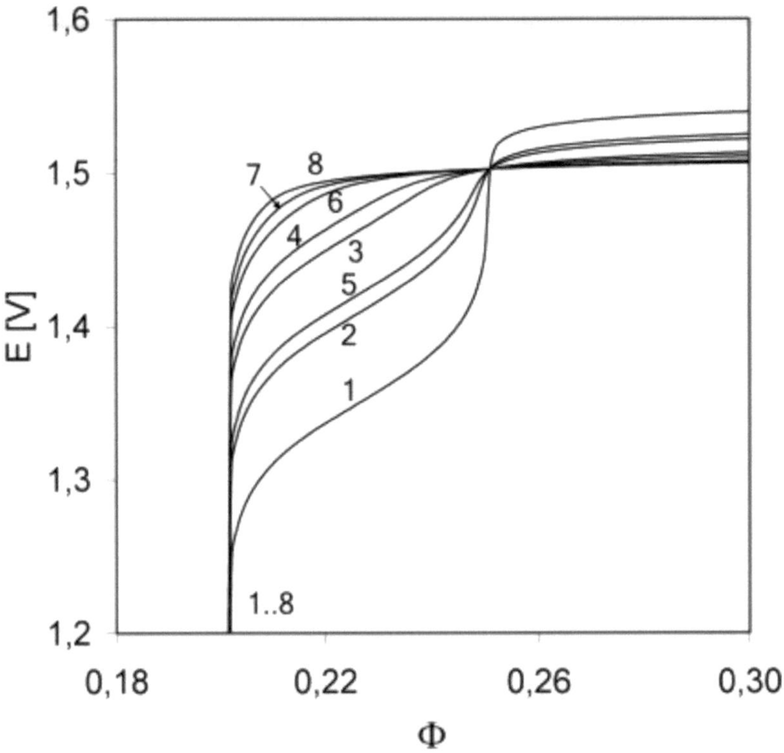

Figure 8. Fragments of hypothetical titration curves for $V_0 = 100$ mL of $FeSO_4$ ($C_0 = 0.01$ mol/L) + H_2SO_4 ($C_{a0} = 0.1$ mol/L) titrated with C = 0.02 mol/L $KMnO_4$, plotted at different pairs of stability constants (K_{31}, K_{32}) of the sulfate complexes Mn $(SO_4)_i^{+3-2i}$: (1) (10^4, 10^7), (2) (10^3, 10^6), (3) ($10^{2.5}$, 10^5), (4) (10^2, 10^4), (5) (10^4, 0), (6) (10^3, 0), (7) (10^2, 0) and (8) (0, 0).

Some equilibrium constants used in calculations may be unknown/doubtful on the stage of collection of equilibrium data. In such instances, the pre-assumed/virtual data can be introduced for comparative purposes, and the effects involved with omission/inclusion of some types of complexes can be checked.

The possible a priori complexes of $Mn(SO_4)_i^{+3-2i}$ are unknown in literature. To check the effect of formation of these complexes on the shape of the titration curve E = E(Φ) in the system (I) ⇒ (II), the pre-assumed stability constants K_{3i} of the complexes, $[Mn(SO_4)_i^{+3-2i}] = K_{3i}[Mn^{3+}][SO_4^{2-}]^i$, specified in legend for **Figure 8**, were applied in the related algorithm, where concentrations of $MnSO_4^{+1}$ and $Mn(SO_4)_2^{-1}$ with the corresponding multipliers were inserted in electron (GEB) and charge balances and in concentration balances for Mn and sulfate. As we see, at higher K_{3i} values (comparable to ones related to $Fe(SO_4)_i^{+3-2i}$ (i=1,2) complexes [39]), the new inflection point appears at Φ = 0.25 and disappears at lower K_{3i} values assumed in the simulating procedure. Comparing the simulated curves with the one obtained experimentally [25], one can conclude that the complexes $Mn(SO_4)_i^{+3-2i}$ do not exist at all or the K_{3i} values are small, when compared with those for $Fe(SO_4)_i^{+3-2i}$ [33].

Other interesting examples involved with 'variations on the subject' are presented in Ref. [1], and other references cited therein.

15. Advantages of GATES and usefulness of chemical processes simulation

Mathematical formalism of electrolytic systems tested and resolved according to GATES principles formulated by Michałowski (1992) arises from synthesis of the three laws: (1°) law of charge conservation, (2°) law of elements conservation and (3°) law of mass action. All other chemical laws result from conjunction of those laws; it particularly refers to the stoichiometry and equivalent mass concepts.

GATES, based on physical, physicochemical and chemical laws, is considered as the best thermodynamic approach to equilibrium, non-equilibrium and metastable, mono- and polyphase, static and dynamic, and redox and non-redox systems, of any degree of complexity, with liquid-liquid extraction systems included [31].

GATES related to redox systems is denoted as GATES/GEB. All these systems are resolvable with use of iterative computer programs, for example, MATLAB. The complexity of chemical systems is here of a secondary importance from the point of view of the computational capabilities inherent in iterative computer programs.

GATES is a confirmation of the thesis that 'everything brilliant is simple'. GATES/GEB is the unique tool to obtain information about the thermodynamics of redox systems on the basis of balances and equilibrium constants values.

GATES enables to simulate all possible (from a thermodynamic point of view) processes obtained after pre-assumed crossing of one or more reaction paths in metastable systems.

GATES enables to simulate the processes impossible to track experimentally; for example, dissolving a solid phase in the electrolytic system of a pre-established composition.

GATES relies on the assumption that the chemistry involved with such systems is predictable on the basis of knowledge of physicochemical properties of the species involved in the system in question. A complete set of non-contradictory relations for the equilibrium constants must be used in calculations; this 'iron rule' of mathematics is then obligatory also in calculations related to electrolytic systems.

GATES is the intrinsically consistent theory, joining fundamental laws of physics and chemistry [10, 28]. The knowledge gaining from redox systems is the most comprehensive way for studying such systems. Note that equations-based simulations are most commonly used in physics and related sciences.

GATES joins, on the thermodynamic basis, four kinds of chemical interactions, named as acid-base, redox, precipitation and complexation reactions, extended on a liquid-liquid extraction in mono- and poly-phase systems. To a certain degree—one can perceive GATES as a spitting image of theory of everything (ToE), as the main, still unresolved issue in physics, aiming to elaborate the consistent theory, that links together four: strong, weak, electromagnetic and gravitational interactions.

GATES referred to electrolytic non-redox and redox systems and is considered as the best thermodynamic approach to such systems. GATES, based on physical (charge conservation), physicochemical (conservation of elements) and chemical (mass action) laws, is the best tool applicable for computer simulation of equilibrium, non-equilibrium and metastable, and mono- and poly-phase electrolytic systems. GATES is the basis for the Generalized Equivalence Mass (GEM) concept, with no relevance to the chemical reaction notation.

One can also express a conviction that the discovery of the Approach II to GEB in context with GATES will lead to gradual elimination of the stoichiometry concept from the consciousness of chemists.

16. Final comments

This chapter provides comprehensive, compatible and consistent knowledge on modeling electrolytic redox and non-redox systems and further steps applied to gain the thermodynamic knowledge on the systems, referred mainly to aqueous media.

The Generalized Electron Balance (GEB) concept, related to electrolytic redox systems, is put in context with the principle of conservation of all elements in electrolytic redox systems, in aqueous, non-aqueous or mixed-solvent media. The GEB is fully compatible with charge and concentration balances, and completes the set of 2 + k equations needed for quantitative description of a redox system, with 2 + k independent/scalar variables $\mathbf{x}^T = (E, pH, pX_1, \ldots, pX_k)$. Two equivalent approaches (I and II) to GEB were proposed (1992, 2006) by Michałowski. The Approach I to GEB is based on a card-game principle, with electron-active elements as gamblers, electron-non-active elements as fans and common pool of electrons introduced by electron-active elements as money. The Approach II to GEB is based on the linear combination 2*f*(O)−*f*(H) of elemental balances: *f*(H) for H and *f*(O) for O. The linear independency/dependency of 2*f*(O)−*f*(H) from charge and other elemental/core balances referred to the system in question provides the general criterion distinguishing between redox and non-redox systems. For non-redox systems, 2*f*(O)−*f*(H) is the linear combination of those balances, that is, it is not a new, independent equation in such systems. In redox systems, 2*f*(O)−*f*(H) is the independent equation, considered as the primary form of GEB and denoted as pr-GEB. The balances for elements/cores ≠ H, O are the basis for k concentration balances, forming—with GEB and charge balance—the set of 2+k independent balances, expressed in terms of concentrations. The Approach I to GEB, considered as the 'short' version of GEB, can be applied if the oxidation numbers for all elements in components forming a system and in the species of the system are known beforehand. The Approach II to GEB needs none prior information on oxidation numbers of all elements in the components and species in the system.

Within the Approaches I and II to GEB, the roles of oxidants and reducers are not ascribed to the components and particular species. The GEB is put in context with the Generalized Approach to Electrolytic Systems (GATES) as GATES/GEB, where all quantitative thermodynamic knowledge on the redox system is involved in the complete set of independent equilibrium constants, where standard potentials E_{0i} are involved. The GATES/GEB provides the best thermodynamic formulation of electrolytic redox systems of any degree of complexity, namely: equilibrium, non-equilibrium and metastable, mono- and poly-phase and static and dynamic electrolytic systems, resolvable with the use of iterative computer programs, applied to the set of nonlinear equations, with no simplifying assumptions needed. The GATES/GEB can also be referred to as redox systems in mixed-solvent media, provided that the related thermodynamic knowledge is attainable. This chapter is referred to dynamic systems, realized according to the titrimetric mode. The results obtained from the calculations can be presented graphically on *2D* or *3D* diagrams. The speciation diagrams obtained according to GATES/GEB have indisputable advantage over Pourbaix predominance diagrams. The GEB concept, unknown before 1992, is perceived as an emanation of the matter/elements conservation, as the general law of nature. The redox systems are formulated on simple principles, unknown in earlier literature. Earlier approaches to electrolytic redox systems, based on stoichiometric principles, are thus invalidated.

Author details

Anna Maria Michałowska-Kaczmarczyk[1], Aneta Spórna-Kucab[2] and Tadeusz Michałowski[2]*

*Address all correspondence to: michanku@poczta.onet.pl

1 Department of Oncology, The University Hospital in Cracow, Cracow, Poland

2 Faculty of Chemical Engineering and Technology, Cracow University of Technology, Warszawska, Cracow, Poland

References

[1] Michałowski T. Application of GATES and MATLAB for resolution of equilibrium, metastable and non-equilibrium electrolytic systems. In: Michałowski T, editor. Applications of MATLAB in Science and Engineering. Rijeka: InTech—Open Access publisher in the fields of Science, Technology and Medicine; 2011. pp. 1–34

[2] Pilarski B, Dobkowska A, Foks H, Michałowski T. Modelling of acid–base equilibria in binary-solvent systems: A comparative study. Talanta. 2010;**80**(3):1073–1080

[3] Michałowski T, Pilarski B, Asuero AG, Dobkowska A, Wybraniec S. Determination of dissociation parameters of weak acids in different media according to the isohydric method. Talanta. 2011;**86**:447–451

[4] Michałowska-Kaczmarczyk AM, Michałowski T. Generalized electron balance for dynamic redox systems in mixed-solvent media. Journal of Analytical Sciences, Methods and Instrumentation. 2014;**4**(4):102–109

[5] Michałowska-Kaczmarczyk AM, Michałowski T. Linear dependence of balances for non-redox electrolytic systems. American Journal of Analytical Chemistry. 2014;**5**:1285–1289

[6] Michałowski T, Pilarski B, Asuero AG, Michałowska-Kaczmarczyk AM. Modeling of acid-base properties in binary-solvent systems. In: Wypych G, editor. Handbook of Solvents. Vol. 1, Properties. Toronto: ChemTec Publishing; 2014. pp. 623–648

[7] Michałowski T, Toporek M, Michałowska-Kaczmarczyk AM, Asuero AG. New trends in studies on electrolytic redox systems. Electrochimica Acta. 2013;**109**:519–531

[8] Michałowski T, Michałowska-Kaczmarczyk AM, Toporek M. Formulation of general criterion distinguishing between non-redox and redox systems. Electrochimica Acta. 2013;**112**:199–211

[9] Michałowska-Kaczmarczyk AM, Michałowski T. Comparative balancing of non-redox and redox electrolytic systems and its consequences. American Journal of Analytical Chemistry. 2013;**4**(10):46–53

[10] Michałowska-Kaczmarczyk AM, Michałowski T. Compact formulation of redox systems according to GATES/GEB principles. Journal of Analytical Sciences, Methods and Instrumentation. 2014;**4**(2):39–45

[11] Michałowska-Kaczmarczyk AM, Toporek M, Michałowski T. Speciation diagrams in dynamic iodide + dichromate system. Electrochimica Acta. 2015;**155**:217–227

[12] Toporek M, Michałowska-Kaczmarczyk AM, Michałowski T. Symproportionation *versus* disproportionation in bromine redox systems. Electrochimica Acta. 2015;**171**:176–187

[13] Michałowska-Kaczmarczyk AM, Asuero AG, Toporek M, Michałowski T. "Why not stoichiometry" versus "Stoichiometry—why not?" part II. GATES in context with redox systems. Critical Reviews in Analytical Chemistry. 2015;**45**(3):240–268

[14] Michałowska-Kaczmarczyk AM, Michałowski T, Toporek M, Asuero AG. "Why not stoichiometry" versus "Stoichiometry—why not?" part III, extension of GATES/GEB on complex dynamic redox systems. Critical Reviews in Analytical Chemistry. 2015;**45**(4):348–366

[15] Michałowska-Kaczmarczyk AM, Michałowski T, Toporek M, Pietrzyk A. Solubility and dissolution in terms of generalized approach to electrolytic systems principles. Journal of Analytical Sciences, Methods and Instrumentation. 2015;**5**:47–58

[16] Michałowska-Kaczmarczyk AM, Michałowski T, Toporek M. Formulation of dynamic redox systems according to GATES/GEB principles. International Journal of Electrochemical Science. 2016;**11**:2560–2578

[17] Meija J, Michałowska-Kaczmarczyk AM, Michałowski T. Redox titration challenge. Analytical and Bioanalytical Chemistry. 2017;**409**(1):11–13

[18] Michałowski T, Michałowska-Kaczmarczyk AM, Meija J. Solution of redox titration challenge. Analytical and Bioanalytical Chemistry. 2017;**409**(17):4113–4115

[19] Michałowski T, Kupiec K, Rymanowski M. Numerical analysis of the Gran methods. A comparative study. Analytica Chimica Acta. 2008;**606**(2):172–183

[20] Ponikvar M, Michałowski T, Kupiec K, Wybraniec S, Rymanowski M. Experimental verification of the modified Gran methods applicable to redox systems. Analytica Chimica Acta. 2008;**628**(2):181–189

[21] Michałowski T, Toporek M, Rymanowski M. Overview on the Gran and other linearization methods applied in titrimetric analyses. Talanta. 2005;**65**(5):1241–1253

[22] Michałowski T, Baterowicz A, Madej A, Kochana J. An extended Gran method and its applicability for simultaneous determination of Fe(II) and Fe(III). Analytica Chimica Acta. 2001;**442**(2):287–293

[23] Michałowski T, Stępak R. Evaluation of the equivalence point in potentiometric titrations with application to traces of chloride. Analytica Chimica Acta. 1985;**172**:207–214

[24] Michałowski T. The generalized approach to electrolytic systems: I. Physicochemical and analytical implications. Critical Reviews in Analytical Chemistry. 2010;**40**(1):2–16

[25] Michałowski T, Pietrzyk A, Ponikvar-Svet M, Rymanowski M. The generalized approach to electrolytic systems: II. The generalized equivalent mass (GEM) concept. Critical Reviews in Analytical Chemistry. 2010;**40**(1):17–29

[26] Asuero AG, Michałowski T. Comprehensive formulation of titration curves referred to complex acid-base systems and its analytical implications. Critical Reviews in Analytical Chemistry. 2011;**41**(2):151–187

[27] Michałowski T, Asuero AG. New approaches in modeling the carbonate alkalinity and total alkalinity. Critical Reviews in Analytical Chemistry. 2012;**42**(3):220–244

[28] Michałowski T, Asuero AG, Ponikvar-Svet M, Toporek M, Pietrzyk A, Rymanowski M. Principles of computer programming applied to simulated pH–static titration of cyanide according to a modified Liebig-Denigès method. Journal of Solution Chemistry. 2012;**41**(7): 1224–1239

[29] Michałowski T. Some remarks on acid-base titration curves. Chemia Analityczna. 1981;**26**: 799–813

[30] Michałowski T, Lesiak A. Euroanalysis VIII Book of Abstracts, PJ 35, PM 10, Edinburgh, Scotland; Young RA. Analyst. 1994;**119**:75N-76N

[31] Michalowski T. Calculation of pH and potential E for bromine aqueous solutions. Journal of Chemical Education. 1994;**71**(7):560–562

[32] Michałowski T, Lesiak A. Acid-base titration curves in disproportionating redox systems. Journal of Chemical Education. 1994;**71**(8):632–636

[33] Michałowski T, Lesiak A. Formulation of generalized equations for redox titration curves. Chemia Analityczna (Warsaw). 1994;**39**:623–637

[34] Michałowski T, Wajda N, Janecki D. A unified quantitative approach to electrolytic systems. Chemia Analityczna (Warsaw). 1996;**41**:667–685

[35] Michałowski T. Calculations in analytical chemistry with elements of computer programming (in Polish). Cracow: PK; 2001.

[36] Michałowski T, Rymanowski M, Pietrzyk A. Non-typical Brönsted's acids and bases. Journal of Chemical Education. 2005;**82**(3):470–472

[37] Michałowska-Kaczmarczyk AM, Michałowski T. General relation valid for electrolytic systems. Journal of Analytical Sciences, Methods and Instrumentation. 2015;**5**:74–85

[38] Sillén LG, Martell AE. Stability Constants of Metal-ion Complexes. London: The Chemical Society; 1964. Suppl. No. 1, 1971 Other options of the notation met in literature: Sillén, L.G., Martell, A.E. Stability Constants of Metal-Ion Complexes, The Chemical Society, London, 1964, Suppl. No.1, 1971.

[39] Lurie J. Handbook of Analytical Chemistry. Moscow: Mir Publishers; 1975

[40] Inczédy J. Analytical Applications of Complex Equilibria. New York: Wiley; 1976

[41] Michałowska-Kaczmarczyk AM, Michałowski T. Solubility product challenge. Analytical and Bioanalytical Chemistry. 2015;**407**(6):1789–1791

[42] Michałowska-Kaczmarczyk AM, Michałowski T. Solution to the solubility product challenge. Analytical and Bioanalytical Chemistry. 2015;**407**(17):4877–4878

[43] Michałowski T, Pietrzyk A. A thermodynamic study of struvite+water system. Talanta. 2006;**68**:594–601

[44] Michałowski T, Pietrzyk A. Quasistatic processes in non-equilibrium two-phase systems with ternary salts: I. struvite + aqueous solution (CO_2 + KOH). Chemia Analityczna (Warsaw). 2008;**53**:33–46

[45] Michałowski T, Borzęcka M, Toporek M, Wybraniec S, Maciukiewicz P, Pietrzyk A. Quasistatic processes in non-equilibrium two-phase systems with ternary salts: II. Dolomite + aqueous media. Chemia Analityczna (Warsaw). 2009;**54**:1203–1217

[46] Michałowski T, Asuero AG. Thermodynamic modelling of dolomite behaviour in aqueous media. Journal of Thermodynamics. 2012;**2012**:12. Article ID 723052

[47] Michałowska-Kaczmarczyk AM, Michałowski T. Calculation of solubility of oxyquinolinates. Journal of Analytical Sciences, Methods and Instrumentation. 2014;**4**(3):71–79

[48] Michałowska-Kaczmarczyk AM, Michałowski T. Evaluation of transition points between different solid phases in aqueous media. Journal of Analytical Sciences, Methods and Instrumentation. 2014;**4**(3):87–94

[49] Michałowska-Kaczmarczyk AM, Michałowski T. Ostwald's dilution law challenge. Analytical and Bioanalytical Chemistry. 2014;**406**(12):2741–2742

[50] Michałowska-Kaczmarczyk AM, Michałowski T. Solution to Ostwald's dilution law challenge. Analytical and Bioanalytical Chemistry. 2014;**406**(22):5251–5251

[51] http://goldbook.iupac.org/C01240.html [accessed on August 5, 2017]

[52] Michałowska-Kaczmarczyk AM, Asuero AG, Michałowski T. "Why not stoichiometry" versus "Stoichiometry—why not?" Part I, general context. Critical Reviews in Analytical Chemistry. 2015;**45**(2):166–188

[53] Beck MT. Chemistry of Complex Equilibria. New York: Van Nostrand Reinhold; 1970

[54] Michałowski T, editor. Applications of MATLAB in Science and Engineering. Published by InTech Janeza Trdine 9, 51000 Rijeka, Croatia, 2011

[55] Katsikis VN, editor. MATLAB—A Fundamental Tool for Scientific Computing and Engineering Applications. Vol. 2. Published by InTech Janeza Trdine 9, 51000 Rijeka, Croatia, 2012